Forbidden Cosmology
Uncovering the Ultimate Connection
(Complete Cosmic Energy Exchange Concept, CCEEC)

G.W. Franke

Wolf Publishing B—Downieville, CA
ISBN: 979-8-218-52443-2
Library of Congress Control Number: 2024920988
Title: *Forbidden Cosmology: Uncovering the Ultimate Connection (Complete Cosmic Energy Exchange Concept, CCEEC)*
Author: G.W. Franke
Digital distribution | 2024
Paperback | 2024

Published in the United States by New Book Authors Publishing

Dedication

To my parents, without whom I wouldn't have seen this wonderful world.

To my many special friends, whose influences have been instrumental in shaping me into the person I am today.

To a few super-loyal pets, including my hero cat, "Ivan," who inspired me to finish this script, your love and support have been a constant source of inspiration.

To this Planet Earth, without which I wouldn't have known the joys and sorrows of life.

To all my readers, I say: "Follow your dreams - they will come true..."

The Author (GWF)

Chapter One
<u>The Current Universe:</u>

We can describe our knowledge about the origin and fate of the universe in five words: "Nothing, Big Bang, Fate Unknown." Yet we had come a long way since the 1920s when a former amateur boxing champion and lawyer turned astronomer named Edwin Hubble made one of the great discoveries in the history of science.

Hubble demonstrated that the spectrum of a moving light source shifts toward the red end of the spectrum if the light source is moving away from the observer. This phenomenon is called the Doppler Effect.

Hubble noticed that the farther away a galaxy is from us, the faster it moves away. He explained that a relationship between a galaxy's distance and speed of recession exists—a distance-to-velocity ratio now known as the "Hubble Constant." It gave us a first glimpse at a dynamic, rapidly expanding universe whose origin and fate had yet to be determined.

By 1929, only 95 years ago, people believed that a mere nine galaxies existed in the universe. The Doppler effect could establish an approximation of their time and distance traveled. Not surprising was the conclusion that we could theoretically reverse the observed trajectories of galaxies and derive at a point in space-time at which the distance between the galaxies must have been zero, a point of infinite density, packed into a speck a million times smaller than the point at the end of this sentence.

We believed for many decades that this point of infinite density exploded somehow from what has been named the Singularity of the "Big Bang." A singularity is a physical phenomenon at which all known laws of physics are invalid, and the law of conservation of energy, which specifies that energy, or its equivalent in mass, can neither be created nor destroyed, was exempted from holding.

Universal law should be valid here, proper there, and faithful everywhere to be objective. There is no reason to believe that energy conservation law cannot be applied consistently in a scientific context, even under extreme circumstances, like the explosion that generated matter phenomena.

All theories about the universe to date—whether the Big Bang Standard Model or the Inflationary model proposed by Alan H. Guth—rely on the Singularity of the Big Bang, which generates a whole array of speculation: Is the universe open? Is it closed? Will it slow down its expansion? Will it come to a halt and contract to its point of origin? Could there be a "Big Crunch"? Could time reverse?

Other essential questions need explanations with today's theories. Stephen W. Hawking asked those in his book "A Brief History of Time" (Bantam Books, 1988):

– "Why was the early universe so hot?"
– "Why is the universe so uniform on a large scale?"
– "Why did the universe start with so nearly a critical rate of expansion that even after nearly 20 billion years, it is still expanding at a nearly critical rate?"
– "What was the origin of early density fluctuations that led to the formation of local irregularities, such as stars and galaxies?

If we could answer all those questions consistently, we would have a *unified theory of nature.*

There are a myriad of old words virtually screaming for new definitions. Gravity is one of them. From Newton to Einstein, gravity's effects have been established and confirmed within the radius of our observable planetary system. Yet the origin of gravity may have to be rethought or redefined.

The gravitational effects in the universe are much stronger than the total amount of mass could account for, initiating a mind-numbing search for so-called "Dark Matter" or "Cold Dark Matter," both of which can not be confirmed or picked up with our high-tech and costly instruments.

We must examine the *electron*'s role as one of the most essential elementary particles in the matter universe. Einstein said, "It would be sufficient to understand the electron!"

The electron was present at the "Big Bang" event. It *is* present in our brains and makes thinking and reasoning possible. The electron regulates heart rates and influences energy patterns within cells. It gives us communication capabilities with computers, television, and radio. Could the electron also be the particle responsible for the expansion of the universe?

This theory gives a new definition of *time*—a concept so little understood that we wreck our brains trying to understand it. Why is time a one-way street? Why does it "flow"? Could time reverse? Why do children perceive a one-hour segment as eternity, while adults see it as nothing but a blip on the ever-faster spinning roulette of life?

Is the *"curvature"* of space attributable to the relative interaction of space, time, mass, and energy in which planets like Earth do *not* orbit the sun due to a force called gravity but rather try to follow a *straight path in curved space* – as Einstein surmised?

What if "curved" space existed by any other means and was unrelated to the presence of matter *in that space*?

Could we explain a force that surrounds us from all sides? A force so strong that its power mystified, even terrified early man? A force our ancestors could not explain, a force from which a belief system arose we now call "religion"?

Could all the elements of "light, heaven, energy, spirit, creation, omnipotence, and eternity" as an element of religion be explained with a scientific universe model that we can confirm but doesn't require "faith" as a condition for existence? Is the "Grand Canyon" between creationists and evolutionists valid, or can they be both right?

Where do we stand?

We stand in a foggy meadow, ten feet away from some railway tracks, watching a fast-moving train emerge and disappear into nowhere. Its departing station and destination are unknown and open to speculation.

Who fired up the train? Who fueled it? Who pulled the lever to move it? Who will apply the breaks to it? When? Is it time to come to a safe halt, or will it smash into a barrier at the end of the tracks at full speed?

We have to move back a mile or two from our current position. Then, we would see the train move between two known stations. All we need is to wait for the fog to burn off. The theory presented here is trying to burn off the fog.

. . .

<u>The Engine of Matter: The Electron</u>

Everything and anything in our universe could not exist without the electron, an elementary matter particle of negative electric charge. Imagine a world without computers, electricity, lightning and thunder, *electronic* data communications, neuron-electric brain activity, or *electromagnetic* fields.

The universe would be dead without the electron.

In principle, it gives us weather, water, and life. "It" is responsible for chemical elements and atomic structure.

We live in an *"electronic"* world: you and me and everything we see and touch. Even the dollar bill we get for our 'labor of love' from employers consists of an outer layer of electrons. We take this atomic structure we call "paper money" to a store to buy a P.C., laptop, iPod, or Tablet to bring the world to us "electronically."

Again, what is the electron?

The electron is an indivisible elementary matter particle. It has a negative electric charge and orbits the nucleus of an atom, which has a positive electric charge. The electron whirls around the nucleus *so fast that its path provides a dense envelope around the nucleus.*

An envelope is a cover to protect its contents!
Protection from what?

Is the electron envelope of the nucleus a mechanism of nature to prevent the nucleus from being *immobilized?* Could it be a mechanism of nature to prevent the nucleus from *repuls*ing a like-charged electric force, for instance?

If so, a like-charged electric force *could* immobilize the nucleus. All of the matter within our universe eventually consists of an electron-envelope over a proton core. Matter could exist as a chemical element and sit in space forever. Yet matter doesn't just sit in space. It is constantly in constant motion, and the universe's expansion has been going on without interruption since its mysterious appearance.

Imagine *what a force it must be* to make billions of galaxies, with billions of stars in each one, move so fast for billions of years without running out of energy.

Could the electron-envelope of matter be responsible for our forward motion and perhaps our motion *toward* a force of opposite electric charge? A force so tremendous, so strongly acting from all sides that the universe expands evenly into space and looks almost the same in all directions?

Could this hypothetical form of opposite electric charge be responsible for the formation of spherical bodies within the vacuum of space, and could we deduce from it that the "roundness" of stars, planets, *and space itself* could be a result of that force instead of ascribing that effect to 'gravity'?

If so, we should seriously search for the possible existence of this force, explain it thoroughly, and ultimately *find* it.

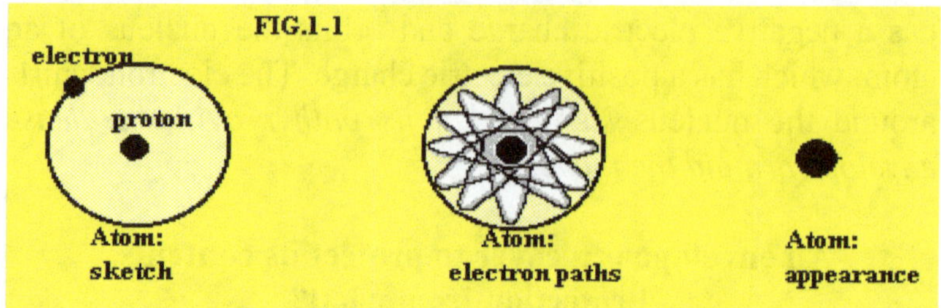

Fig.1–1 shows the principle structure of all matter: an electric-positively charged proton (core) is protected from eventual harm by the protective paths of the electric-negatively charged electron.

The exclusive electron-envelope-innate to all ordinary matter – gives matter a surface net-electric-negative charge (*).

If electrons are attracted to a hypothetical form of opposite electric charge, they will move the nucleus within the atom at the same speed. This mechanism would guarantee the integrity of matter, and the complete atom (star, planet, galaxy, galactic cluster) would arrive at its destination in one piece and simultaneously.

*Matter's net-electric negative charge arises because electrons can easily separate from natural and artificial surfaces and compounds by friction, motion, heating, energy conversions, and explosive disintegration. Free and unattached electrons are a common occurrence in nature. Example: a side effect of human food consumption is the generation of free electrons – free radicals that bounce inside cells and damage their structure over time, leading to premature aging. Another example is solar flares, which can generate millions of streams of electrons that interfere with electronic communication on Earth. The abundance of stars in the universe and their electron emissions provide the "electron-mass" on which a yet-hypothetical form of opposite electric charge can act to further the expansion of the electronic universe in an unhindered way.

Current models of the universe need to see a connection between the universe's speed of expansion and the electron envelope of matter. They also fail to see and cannot explain where the energies of the "Big Bang" originated. *Should* we ignore the possibility of such an *originating* force or attempt to find it? American science has chosen ignorance, confusion, and ridicule over progress and open-mindedness.

The question is this: *What force have we chosen to ignore?*

. . .

Searching for an Unknown Force

We know from experience that all things in our universe come either in pairs or in opposites:

Male – Female
Cold – Hot
Black – White
Negative – Positive
Matter – Antimatter

We conditioned ourselves to believe that the universe somehow "created" these opposites to function. It is the same as a lie repeated a thousand times becomes the truth, and no further investigation is needed. The most important question to ask is: *Why?*

–Why would a cold, dark vacuum and its matter exist as the unilateral force of cosmic nature without having an opposite (counterpart)? After all, observable evidence confirms the duality of nearly everything in the entire universe.
–Why should space be "curved" when the mass that should be there to achieve that effect is not there?

Could the answer to these questions come from an "antimatter" universe?

Stephen W. Hawking also fell into the trap of looking at the black bubble in which we are anchored from *within,* claiming that large quantities of antimatter *within* our universe cannot exist "because matter and antimatter would attract each other and annihilate which would give off high-energy radiation which we could see and measure."

First, consider the mystery of cosmic microwave background radiation and ultra-energetic gamma-ray bursts. We know that Gamma-ray bursts do explode literally out of nothing and in regions of space not containing any observable matter. The destructive power of these rays, however, is well known from experiments and the detonation of nuclear bombs. Something *has got to be there to generate gamma rays.*

The accepted theory maintains that Gamma rays are ordinary matter's highest-frequency/shortest-wavelength manifestations; however, this is only partially true. Extraordinary matter manifests itself a dash beyond the capability of existing instruments. How this works out will be shown later.

Should it be determined that antimatter is responsible for these occurrences, we would need a new understanding of this process. We know that large amounts of antimatter don't exist in the observable universe. We are still determining, however, how much or how little volume in antimatter is needed to bring about galactic destruction.

The problem lies in imagining the possible qualities of antimatter. No antimatter- planets or galaxies exist, and antimatter is *not* a simple reversal of matter with an opposite electric charge.

Important in Stephen W. Hawking's admission is the fact that matter and antimatter would "attract each other."

What could antimatter be?

The "anti" of a cold vacuum is a hot "something"; perhaps a super-energy plasma.

The "anti" of limited cold could be unlimited heat.

The "anti" of a negatively charged electron? – A positively charged electron, which exists and is called POSITRON!

The "anti" of atomic matter would be "non-atomic antimatter." Non-atomic antimatter equals "elementary antiparticles."

The "anti" of black would be white bright/sunny.

The "anti" of finite space would be "infinite antispace" and so on.

From these statements, we could extrapolate the possible qualities of antimatter:

It is pure elementary antiparticle energy of infinite heat and infinite size, with a reversed electric charge to matter, not suspended in a vacuum, which attracts matter.

Upon contact of the two, the matter is disassembled into its elementary particles, forced to reverse its temporary electric charge back to the original charge, and becomes the property of the "positronic" energy again.

If we also can envision the incredible energies released from that contact, this is a region of extreme danger to the survival of antimatter *in that region.*

The universal "bubble," which contains no particles or heat energy after the contact, is *right in place* to put a wedge into this overheated region.

This separation of areas by universal vacuums is the cosmic built-in mechanism to prevent a region from self-destruction. At the same time, it provides the necessary physical contact to perpetuate the larger quantity and quality infinitely. It could be the equivalent of a rejuvenation of antimatter-energies without exaggeration—a life without death. It has always existed. "Eternity" – if you will.

What we would need is an approximate understanding of this mechanism insofar as it must be consistent with the law of conservation of energy (better yet: the law of conservation of energies; plural) without resorting to the excuse of a "singularity without symmetry," or a "Big Bang" out of nothing.

. . .

Chapter Two
Reason for a More Complete Universal Model:

Why should two separate universes – the dualverse – be more correct than a single universe?

Quite simple: back in the 1920s, when less than ten galaxies were known to exist, we could agree that "perhaps, maybe, possibly, who-knows-better" all of these galaxies were once so close together that they could have made a point of infinite density which "somehow" exploded in the Big Bang.

Over the years, we have discovered hundreds, thousands, millions, and billions of galaxies. We have yet to stop counting.

In October of 1990, the American Journal "Science" published an article by J. R. Kuhn, Michigan State University; Juan M. Uson, National Radio Astronomy Observatory, and S.P. Boughn, Haverford College, about the discovery of a concentration of galaxies stretching for more than 6 million light years. Up to then, it was the most significant cluster discovered. The central galaxy of the galactic cluster, named "ABELL 2029," contains more than one hundred trillion stars (trillion with T).

"ABELL 2029" is only one of many clusters already known or yet to be discovered. It makes the notion of a "point-origin" of our universe and a Big Bang explosion out of nothing more and more unlikely. Not only do we look for and discover ever more galaxies, but we also look at their shape, spectral light, and make-up to determine their possible age. In 1988, a "mileage record" was broken when a team of scientists announced at a meeting of the International Union in Baltimore that they had discovered the "most distant galaxy yet seen by man."

Named 4C41.17, it is some fifteen billion light years away from us. It calls into question the estimated age of the universe itself. Although 4C41.17 is more lumpy, elongated, and more turbulent than most other galaxies, its plain existence just one or two billion years after the Big Bang implied severe problems with current cosmological models. Developing complex and complete galaxies in current theories should have taken much longer.

Either the universe is considerably older than we suspect, or the formation of galaxies was accomplished much faster and by processes unknown to us to date.

The "Complete Cosmic Energy Exchange Concept, CCEEC" will try to explain a different approach to galaxy formation.

The idea of a larger universe should have emerged much earlier from the work done on the electron. The German-born physicist Hans Dehmelt devoted much of his life to studying the electron (Nobel Prize). He was obsessed with trapping a single electron to study its spin and magnetism—two of the four characteristics defining an electron. Mass and charge are the other two.

From 1959 on, Dehmelt experimented with several approaches, employing what has come to be known as a "Penning Trap." Finally, in 1973, he trapped the first single electron in an almost perfect vacuum without letting it escape. How did he do this?

Dehmelt based his device on an electric component called the Penning discharge tube. Electrons leaped into the tube from a hot-tip wire inserted through a hole in one cap. Dehmelt placed the trap inside a magnetic field that pointed straight up and down along the length of the tube.

The vertical magnetic field prevented the electron from flying horizontally into the trap's walls. A charged particle trying to cross a magnetic field is always pushed sideways, at right angles, to both the field and the particle's forward motion. A strong magnetic field pushes so hard sideways that it forces the electron into tight, circular horizontal orbits.

In addition, a negative electric charge on the two end caps prevented the electron's vertical escape, forcing it back up toward the top cap and repelling it again. The single electron moved in a tight circular motion at the center of the Penning discharge tube, or "Penning Trap."

FIG.2–1

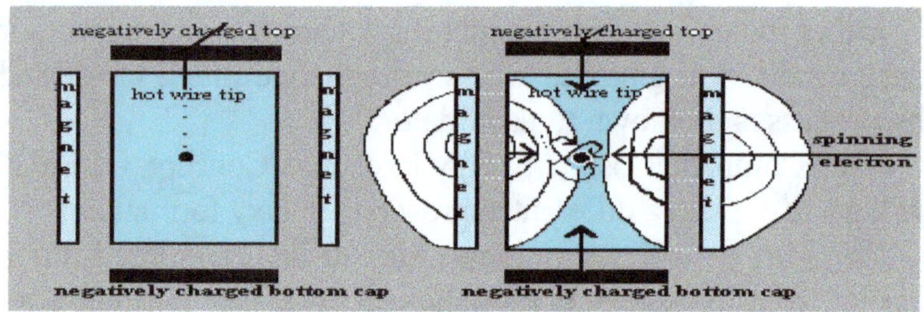

In 1989, Hans Dehmelt shared the Nobel Prize in Physics for his groundbreaking work on the characteristics of the electron.

Unfortunately, Dehmelt didn't see a connection between the "Penning Trap" principle and a possible dualversal model. Conditioned by Big Bang theorists, he believes in a particle called Cosmon, "the heaviest particle ever to appear, the particle that gave birth to the universe."

But take a second look at Dehmelt's contraption (Fig.2–1). If you round off the corners in the sketch, imagine a three-dimensional vacuum, and exchange the label "electron" with "Cosmon," you almost get the miniaturized version of the universe in a bottle. The duality of the physical universe becomes visible.

A particle propels itself into a vacuum. A combination of two forces of nature – a strong positron-magnetic field and direct repulsion using a like charge – holds the particle in a fixed position. It can not escape *unless it reverses its charge or the charge of the surrounding wall reverses*. Motion initiates in the direction of the charge difference. A charge *opposition*, then, is the most efficient way of universal nature to force particles over long distances to maintain a predetermined, even increasing speed until such time in which a merger with the originating energy can erase the charge difference.

Einstein once said: "Explain the universe. Be brief." This theory is as brief and straightforward as possible without leaving any gaps.

Please pay particular attention to the following few pages. They provide the most fundamental and necessary information from which all other facets of life become clearer: religion, politics, psychology, the state of consciousness, and your position amidst it all.

. . .

The Dualversal Mechanism:

Complete Cosmic Energy Exchange Concept, CCEEC:

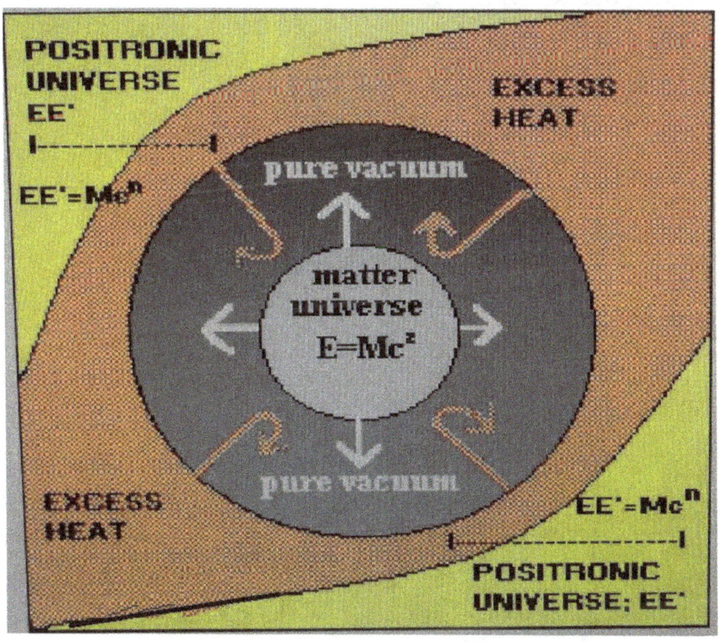

Chapter Three
<u>The Dualverse:</u>

<u>Complete Cosmic Energy Exchange Concept, CCEEC,</u>

Any cosmological theory should assume the initial conditions of the universe and its energies that agree with observations and generally accepted laws of physics.

The CCEEC postulates a complete cosmic energy exchange concept that explains the existence of an infinite universe of opposite electric charges. It consists of a plasma of yellowish/bright energy whose most remarkable feature is its homogenous texture.

FIG.3A

With magnification, the plasma would exhibit irregularities that emerge randomly throughout the energy as tiny black dots of various sizes:

FIG.3B:

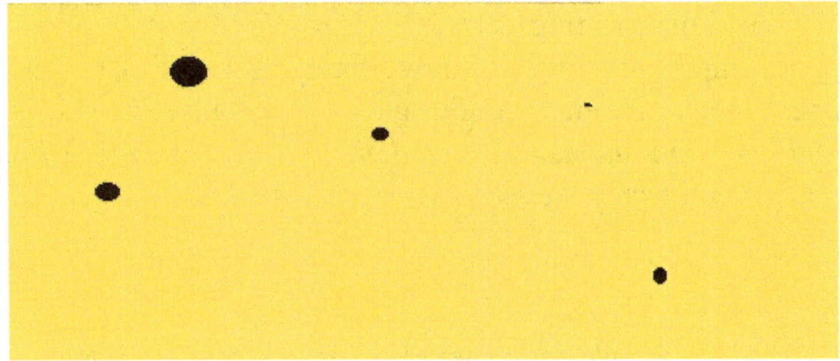

Further magnification would reveal that the distribution of 'dots' is not random. Dots appear to be congregating in areas of extreme heat and differing brightness. Regardless of what area you choose, any closer magnification will reveal the same principle—each "dot" in the infinity of the energy is either a particle-free vacuum that can become the embryo of a "matter universe" or is already an "electronic" matter universe in various stages of development.

Following is a step-by-step account of these mechanics and original energies. It shows the formation of a matter-universe, its development, and the final utilization of its energies in the future.

The cross-section: a yellowish-white sky with "black stars."

FIG.3C:

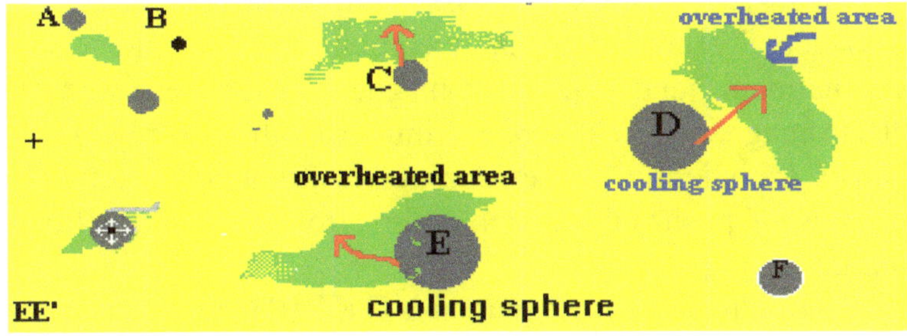

An energy plasma consisting of elementary antiparticles

exists in which "heat-free," particle-free cooling spheres exist as a natural mechanism to separate and displace areas of occasional overheating.

Elementary antiparticles move freely and so fast that they appear stationary in a yellow glow whose visible spectrum *would be continuous without omission or absorption lines because no chemical elements are present.*

. . .

The Working Model of the Dualverse:

Fig.3–1 below shows a further magnification of cooling sphere "E" of the previous page; see Fig.3C

The matter-filled universe expands from inside out toward the edge of a finite-size "cooling sphere" or vacuum, part of the–surrounding elementary antiparticle energy. The void into which we expand has two functions:

A) To allow unhindered expansion of matter energy at and below the speed of light. Equation: $E=Mc2$

B) To insulate matter from contact with antimatter long enough to allow the development of life and intelligence.

The "electronic" matter universe is evenly attracted by and to the positronic universe. This attraction is responsible for the steady and unstoppable speed of our expansion some 15-20 billion years after the emergence of matter.

The so-called "missing" mass of the matter universe could now actually be explained. It combines the supplementary gravitational and "positron-magnetic" (as opposed to electromagnetic) field forces from that other energy. Those values are slightly off-key for our instruments because the physics of positronic matter differs from that of electronic matter.

The positronic energy **equation is E.E.' = Mc*** (*=n, infinite.)

Fig. 3–1:

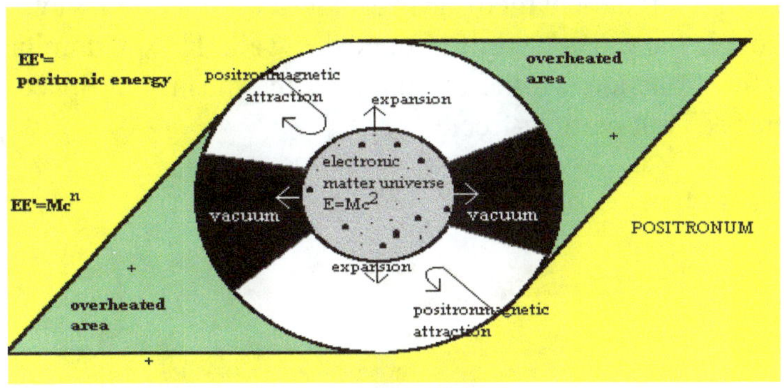

. . .

<u>The 'Seeding' of a Universal Vacuum:</u>

Close-up of void "E" from FIG.3C,

FIG.3–3:

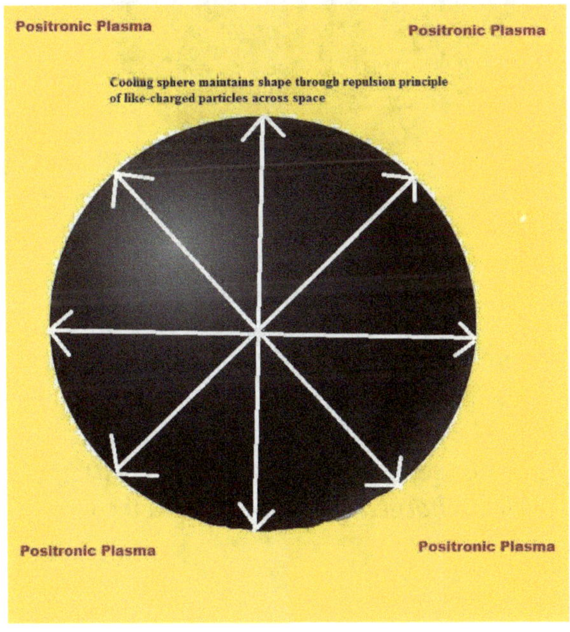

The cooling sphere maintains its shape due to the repulsion principle of like-charged particles that act from all sides evenly through the vacuum on the opposite wall. The predominant particle outside of the vacuum is the positron. Space is "curved" but particle-free.

FIG. 3–4

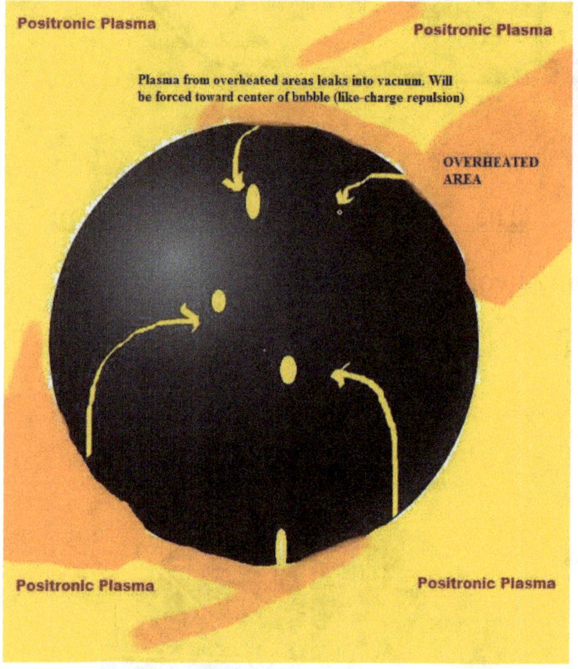

Plasma areas occasionally overheat, allowing particle-free cooling spheres to move into these areas for heat dispersion slowly. Some plasma spills into the vacuum. The repulsion principle of like-charged particles prevents the spill from reentering. The separated energy *forces itself (repulsion principle of like-charged* energies) toward the center.

Fig 3–5

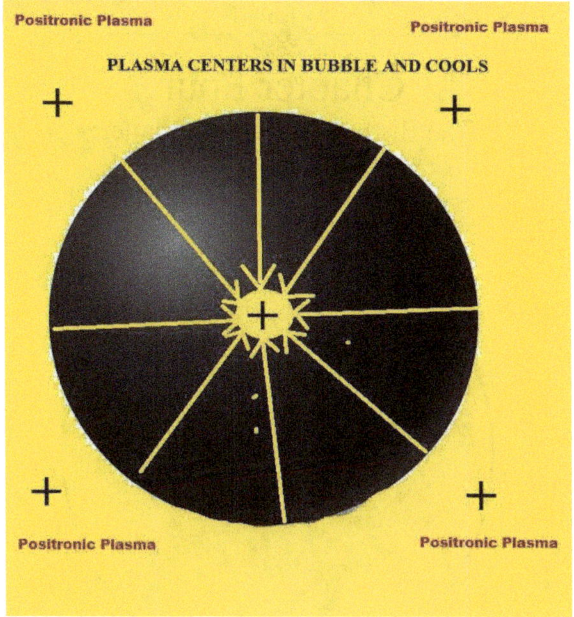

The center of the vacuum begins to spin in a circular motion without a chance of escaping.

This "blob" of energy could have a radius of several million light years. The following illustrations will show this "blob" and how it converts itself to matter.

. . .

Chapter Four
The Birth of Electronic Matter:

FIG.4–1

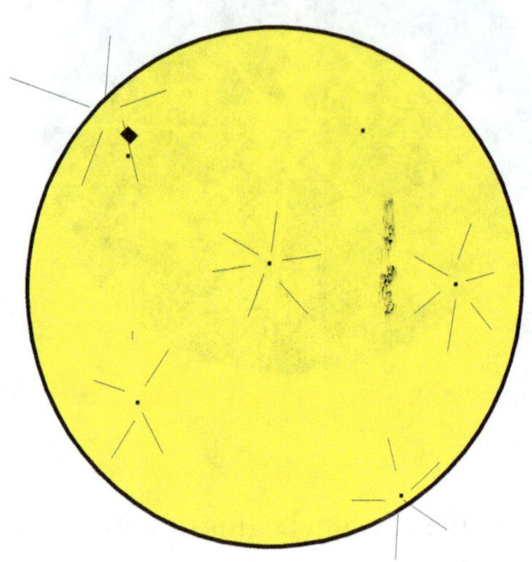

The positronic plasma spins inside the vacuum. The cutoff from the mother-energy prevents it from replenishing its energies. It radiates its now limited energies into space and slowly "cools."

Like bacteria attack an orange left to rot on its *weakest* surface area, positronic energy is attacked by an electric-charge change of some elementary antiparticles, which become the new quality of the electron.

Electrons are short-lived and annihilated.

FIG.4–2 cooling

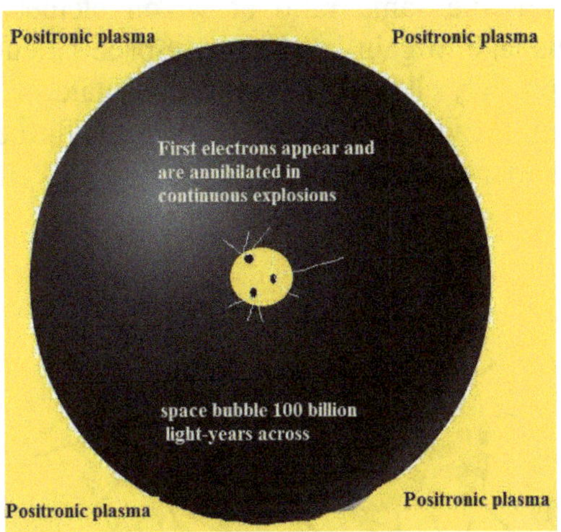

More "cooling" allows the "on-again" and "off-again" the existence of moving elementary matter particles, who can now briefly exist as separate entities in the weightlessness of already existing space. Matter particles' forward/backward motion establishes a "time" reference for their existence.

Temperatures are in the trillions of degrees.

FIG.4–3:

As time passes, the energy of the positronic plasma diminishes at the same ratio at which elementary matter particles manage to gain in quantity. Space (vacuum) allows the two oppositely charged qualities to insulate space between them. Intense radiation energy leads to further cooling. Escaping matter-energy is attracted back to the center.

FIG.4–4

Continuous radiation significantly compromises the purity of the original plasma. Larger sections of electronic matter (electrons and quarks) can exist for extended periods, and a pattern takes shape.

FIG.4–5

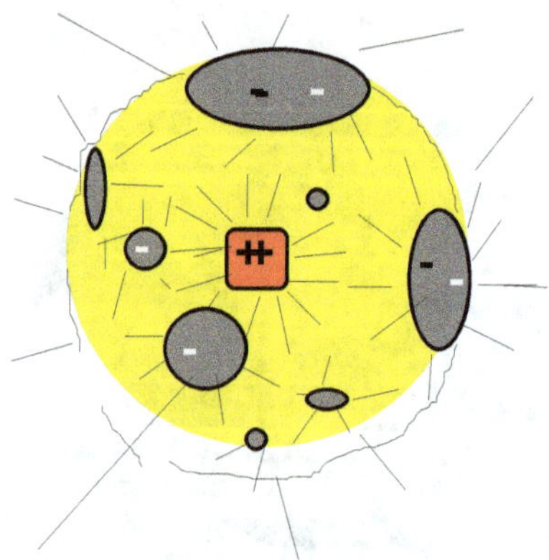

A pattern emerges when matter-energy begins to gain the upper hand. The abundance of the *electron* in the new energy makes it sway away from equilibrium: its combined quantity wants to follow its natural attraction toward the opposite charge coming from the *wall* of the cooling sphere, but the much closer remaining positronic energy has the "first say" in attracting the newly formed, oppositely charged quality and quantity.

FIG.4–6:

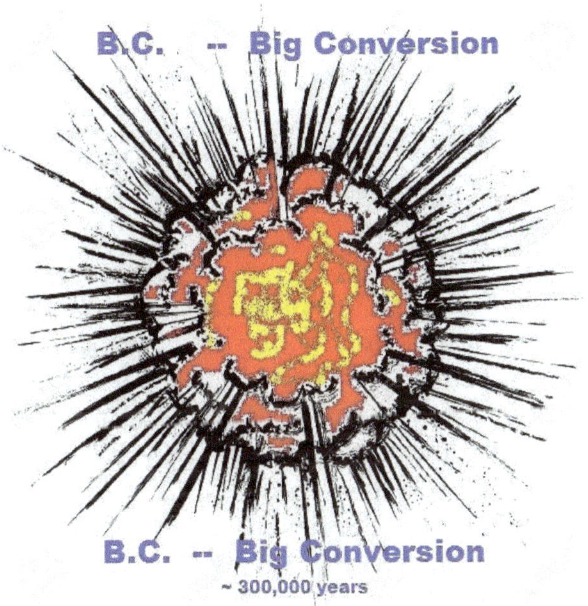

With the sudden strength of a mortally wounded soldier who takes just one more enemy soldier with him as he falls, the plasma attracts.

The oppositely charged energy buries it in a gigantic explosion a trillion times more potent than the brightest, hottest supernova ever seen by man: the so-called Big Bang, or new: the **Big Conversion.**

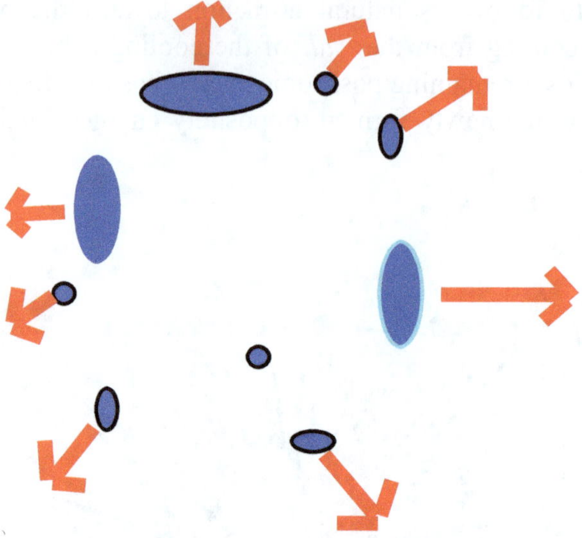

The result of the Big Conversion (B.C.) is a clean slate. The plasma disappears. The exclusive electric charge of elementary matter particles makes it instantly attractive to the all-encompassing positronic energy of far away. The expansion of the matter universe begins (11.2 million miles per minute).

The pattern established before the completion of B.C. reemerges is the reason for regions with superclusters and areas with few clusters in the universe.

. . .

Chapter Five
The Fate of the Matter Universe in Finite-Sized Space:

FIG.5–1

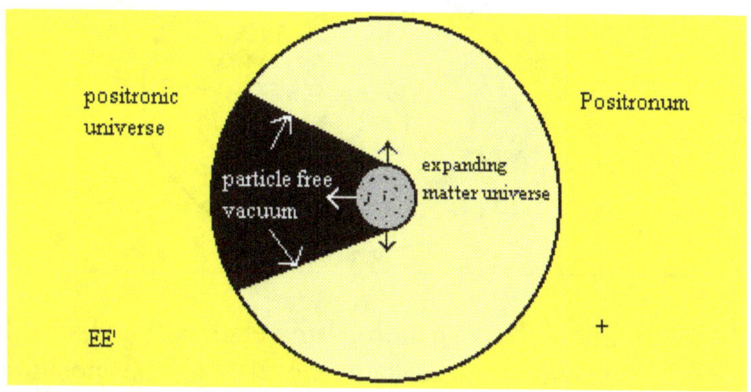

The Big Conversion explosion generated elementary matter-energy, which later coalesced into stars and galaxies. The word "matter universe" only applies to the particle-filled part that extends into the void toward the positronic plasma.

FIG.5–2:

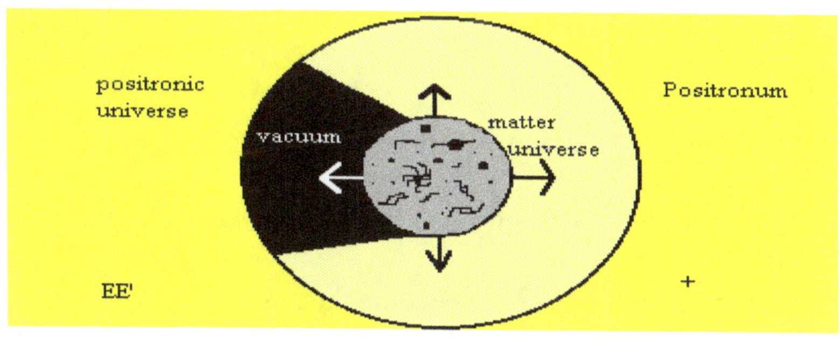

The universe's "edge" expands at the speed of light (11.2 million miles per minute). The closer a galaxy is to the outside of the universal bubble, the faster it moves. This distance/velocity ratio is called the "Hubble Constant."

FIG.5—3

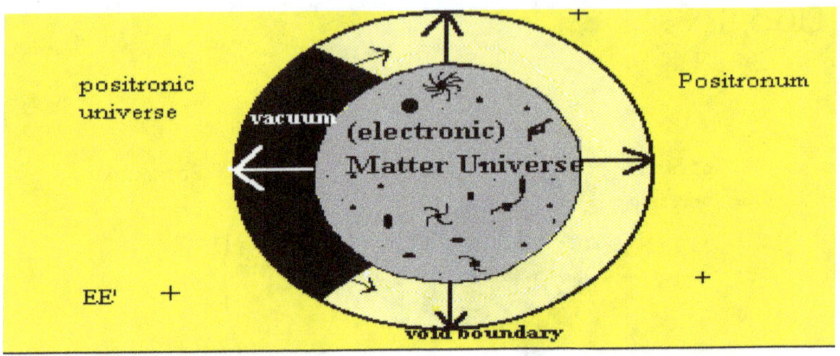

In the future, the expanding universe will exhaust the available insulating space between it and the positronic plasma. Upon contact, matter will revert to elementary particles and change its electric charge to the original, again becoming elementary antiparticles.

As such, matter will *not be annihilated* or perish forever. Its combined energy returns the lost energies of the plasma and rejuvenates it for another cycle.

Developed intelligence becomes part of the plasma. Think about this.

. . .

From Energy to Matter:

A New Theory of Element Formation:

This new cosmology could become the basis for new scientific research and school curricula directions. The reason: It draws a much wider net and brings together many current and seemingly unrelated ends in physics, astronomy, cosmology, and philosophy. It also constitutes a more complete accounting for the energies required to derive our current state of existence.

This scenario eloquently demonstrates that the so-called Big Bang must not necessarily be an exclusive starting point for a matter universe. Many already developed theories dealing with events after the Big Bang would not have to be changed drastically.

The Big Conversion explosion remains the watershed transition point that transformed all original energy within *our vacuum* to a new quality from which biological intelligence could arise to contemplate it.

The supernova of the Big Conversion explosion had such superior energy, power, and heat compared with anything matter could later achieve in nuclear fusion that the three forces of nature—the electromagnetic, the weak nuclear, and the strong nuclear force—*were all the same.*

More logical would be that a field of *elementary matter particles*, so hot that particles couldn't cling to each other, could not have had a weak or strong nuclear force since atoms hadn't formed yet. Even gravity did not exist before, at, and right after the Big Conversion explosion because everything happened in the weightlessness of already existing space that was 'curved' *before* any energy particles existed.

The Big Conversion explosion spewed out a myriad of elementary particles that *initially* consisted mainly of two particles: anti-positrons and "anti"-antiquarks, *which have been named Electrons and Quarks* by physicists of our universe.

In the weightlessness of cold space, the quarks were quickly shocked into the formation of triplets: two up-quarks and one down-quark, which became electric—positively charged Protons.

These triplets of quarks, protons, could only survive in a nucleon gas formation if they had a proportional number of electrons mingled in between to protect the protons from the repulsive, electric-like-charged force to which the electron was attracted.

To get Hydrogen, one triplet of quarks with a net-positive electric charge has to attract one electron with a net negative electric charge to assemble the simplest and most abundant building block of matter. Since quarks and electrons were the initial elementary matter particles after the *Big Conversion* explosion, they also had the best chance to unite before other, more complicated forms of matter could do the same.

I state that practically *all* matter atoms and natural chemical elements were formed simultaneously within minutes of the *B.C.* explosion. This step has already been confirmed by observations using the Hubble Space Telescope.

By analyzing space telescope data, Jason Cardelli of the University of Wisconsin at Madison found traces of heavy elements – lead and thallium – whose nuclei contain 82 and 81 protons in a giant gas cloud 400 light years from Earth. {Discover Magazine, 5/94, page 20}.

Today's accepted theories maintain that all elements outside Hydrogen, Helium, and Lithium are forged inside stars or in supernova explosions. Could scientific dogma be more inconsistent than that? The Big Conversion Explosion, or even the Big Bang by itself, *is the most giant supernova ever to appear* in our space. Such an event should, therefore, be much more efficient in forming all known elements than any subsequent more miniature supernova. Once formed, these elements develop a "material memory," which makes them an innate part of the physics and chemistry of stars and galaxies. Galactic or stellar explosions may temporarily disassemble these elements into elementary particles. We interpret their subsequent reconstruction after those explosions as a new "element creation."

. . .

Chapter Six
Proton to Neutron Flip:

Imagine the pure and superior energies of the *Big Conversion* explosion. The Energy spilled into existing space at high speed. Particles slowed down due to radiation (heat loss/cooling) and expansion into the void. As particles seek to hook up with members of the opposite sex – pardon me – the opposite electric charge, they get closer together and reduce the space between themselves. The reduction of space between particles generates a different medium – a more compacted state of existence that emerges as a visible division of matter against an otherwise empty vacuum.

Protons and electrons collide, repulse, and attract each other according to their electric charge.

If two protons collide, they repulse each other due to their like-electric charge. The same happens to two electrons in a collision. If, however, two protons collide *the exact moment an electron joins the two, one of the protons is forced to "flip" the internal electric charge of its quarks to prevent repulsion from the other proton. This way, one of the protons loses its repulsive force and becomes neutral. In modern physics, we call this particle naturally a Neutron,* and the atom described is deuterium, an isotope of hydrogen.

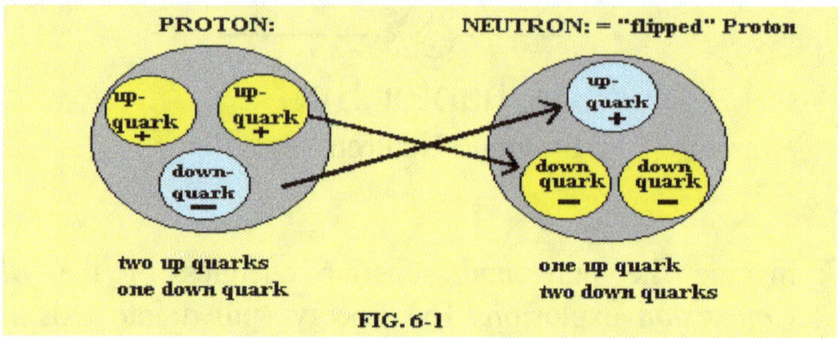

The configuration of the proton's quarks must produce a net positive electric charge (for the attraction of an electron), while the configuration of the Neutron must be electrically neutral. Current theories give the Up- and Down-quarks two-thirds of a positive charge *each*, while the Down-Quark has an electric charge of 1/3 negative.

Proton: 2/3+ + 2/3+ +(1/3-) = 1+   NEUTRON: 2/3+ +(2/3-) = 0

We should accept the notion of a forced "flip" of electric charge in single atoms or particles as a scientific fact to settle once we have the technology to take pictures of the charges of individual atoms.

To date, this has not been possible. The deductive evidence comes from research done on static surface electricity. Surfaces become charged because rubbing brings the atoms of both materials close together so they can interact. Electrons are torn away from one surface and leap to the other. When the two surfaces are separated, they have opposite charges.

. . .

Chapter Seven

Other elements: Deuterium, an isotope of hydrogen, is also called heavy hydrogen because its nucleus contains one proton and one Neutron underneath the single-electron envelope. This light element was predominantly present in the early universe and is a direct result of the energies of the BC explosion. Right after the BC explosion, a single electron could capture two protons: temperatures were still extremely high, which allowed the forced electric "flip" of a proton to make it neutral. We theorize that all neutrons formed approximately simultaneously and represent a conversion product of an existing elementary particle cluster (quarks).

It is also plausible now that other chance combinations of chemical elements could have formed the very same way, including the formation of heavy elements whose quantity depended only on the immediate availability of free electrons. It concerns only about two percent of all the matter in the universe—that part of the universe *does not consist of hydrogen or helium*. 98% of ordinary matter consists of hydrogen or helium and their isotopes.

Why is it so?

In the dualversal model of this theory, positively charged matter, such as quarks or protons, could not have expanded into outer space, which is under the long-range influence of a positively charged other form of energy. *If electrons had not joined immediately with protons at the first chance, matter as we know it would not have developed.* Clouds of negatively charged electrons would have moved toward the boundary of our cosmic cooling sphere at the speed of light. At the same time, protons would spin exclusively at the dead center of this universal bubble, neither being able to form stars or galaxies nor ignite nuclear fusion.

How does the CCEEC theory of proton, neutron, and element formation compare with the current theory of Big Bang Nucleosynthesis and the predicted and observed abundance of light chemical elements, from hydrogen to Carbon?

In principle, there is no disagreement about the numbers. Big Bang Nucleosynthesis asserts that matter began to merge when the ratio of neutrons to protons was one to seven and that virtually all neutrons were swept up early for the "creation" of helium-4 nuclei, which account for their early abundance. Then, "the creation of other elements stopped...because no stable nuclei emerge when a helium-4 nucleus interacts with a proton, a neutron, or another helium-4 nuclei. Most other elements emerge within stars, which have sufficient densities to compress helium-4 into heavier elements."

The Big Bang Nucleosynthesis theory could have a fatal flaw. It must stay within the framework of Einstein's gravity definition, which asserts a linear formation of heavy elements using gravitational compression.

The opposite is true: elements can only be assembled and formed at or right after gigantic explosions in which matter disintegrates into its constituent elementary particles. Matter is built up from scratch everywhere, but it arises only according to its internal material memory, composed at the Big Conversion explosion.

As we will be able to look deeper and deeper into space, and thus closer to the energies of the BC explosion, we will find in the optical spectrum the emission lines of most of the known chemical elements.

It would also help to agree on the term "cosmic energy":

Energy is not an undefined "soup" of yet-to-be-classified hot nothingness from which elementary particles spring somehow.

In the conversion process of the CCEEC model, energy in its pure state is *the free, non-attractive motion of elementary antiparticles with an electric charge: positive and negative. These particles changed their electric charge to the opposite when a prolonged deprivation of sustaining energies occurs within cold space.* (Radiation "cools" energy). The context here is that the lowest possible temperature of positronic particle energy *is* the *highest* possible temperature of matter particle energy. Both cannot occupy space simultaneously for prolonged periods, which is paramount to the strict biological division between life and "death": the last breath of an organism is the necessary prerequisite to 'see' a different dimension.

The following figure shows the CCEEC Principle of simultaneous element formation out of elementary particle energy:

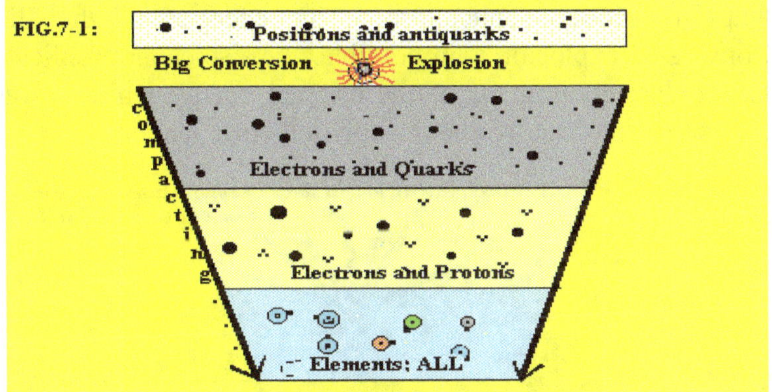

After the BC (Big Conversion) explosion, elementary matter-energy (quarks and electrons) condenses to protons and electrons in nucleon gas formations. Radiation "cools" the new energy, and electrons start to capture the first protons, accidentally forming neutrons.

Fig.7–1 above shows an abundance of elementary matter particles at and after BC (as shown in the upper third) 'compacts' to completed chemical elements and atoms in the lower third. It is obvious why hydrogen and helium are more abundant than other atoms: it takes less energy for an electron to hook up with a single proton or two than to construct elaborate elements.

Chemical elements past Carbon-12 are the reluctant chance by-product of element formation. They account for less than two percent of the Mass in the universe.

It underscores again the point that more *meaningful* exotic matter families might not exist. If they do, they are irrelevant in the construction and evolution of ordinary matter. The apparent short-lived formation of exotic particles in high-speed particle accelerators may be the resonance of a "wounded" proton colliding with another proton, and we wrongly interpret these resonances as separate particles.

However, for decades, $50 billion Accelerators and millions in annual salaries have provided a safe living for thousands of scientists and their families.

At this point in our technological development, it is clear that an accurate replication of the energies of the Big Conversion explosion is impossible. We also cannot replicate the gradual cooling off of the original energy while it was expanding into an already existing universal vacuum.

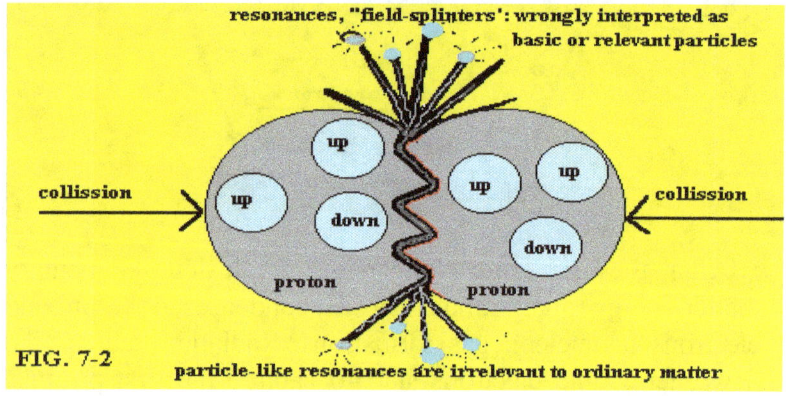

FIG. 7-2

. . .

<u>From Nucleon Gas to Energy Spheres:</u>

How would a nucleon gas cloud organize if it entered the chill of a pre-existing vacuum of immense dimensions? It could be as much as 100 Billion Lightyears across. Wouldn't it try to 'curl up' according to the strength of its internal electric charge differential and maintain the highest energy efficiency as a sphere?

Much like air, when released under water, will form spherical bubbles, or soap bubbles, created from the surface tension of its atomic chemical structure against the lighter medium of air, a nucleon gas formation will also form spherical bodies because the sum of its elementary particles *has substance* against the non-substance of the surrounding vacuum. It so preserves its energy in space.

At first, nucleon gas clouds of any size start forming elliptical, then spherical shapes. Moving away from the BC explosion, the hotter-than-nuclear-fusion-gas is quickly subdivided into millions and millions of smaller energy spheres. Their round shape is partly a result of the positron-magnetic attraction of the all-surrounding energy far away.

Each of these primordial spheres may become a galaxy or galactic cluster. With time and distance, the spheres increase the space between themselves while they are "attracted away" into outer space. The subdivision of a nucleon cloud into energy spheres may develop as shown:

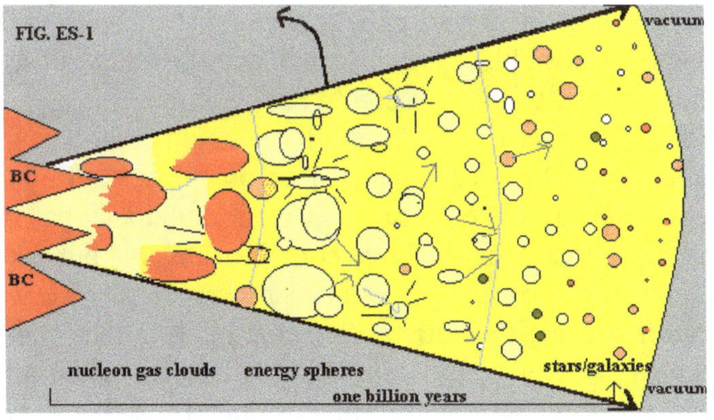

The countless different-sized "energy spheres" in the early universe are still colliding with each other, merging, and dividing until they garnered enough room and speed to make these mergers and collisions less frequent.

FIG.ES–1 shows the principle development of ancient galaxies, which might later develop into galactic clusters of any proportions. Their size is only dependent on the total number of available elementary particles.

Galaxies of this epoch had yet to reach their final shape as they needed a few hundred million more years to organize themselves under the influence of their rotation in conjunction with their simultaneous expansion into space. Their shape at the time could have been irregular and elongated, just as the 1988 discovery of 15 billion years old galaxy 4C41.17 in the beginning chapters of this theory.

· · ·

The "Honeycomb" Structure of the Universe:

A few years ago, PBS television presented a series on astrophysicists' mapping of galaxies. They made a startling discovery that no one has been able to explain: galaxy clusters arranged themselves in a "honeycomb" structure, and the matter itself is displayed as if it is outside of huge galactic bubbles. These bubbles are so giant that their walls can stretch for hundreds of millions of light years.

Nobody knows *what gigantic* force could exist to explain this mystery. The preceding chapter on element origination explains the principle of nucleon cloud subdivisions. If we look not only at a small segment of the universe but take all the energy spheres and place them in perspective, we would see a whole conglomeration of cosmic bubbles that make up our known universe. But something still needs to be added. That "something" appears as soon as I put a sketch of a heavy atom next to the universal bubble sketch, as shown:

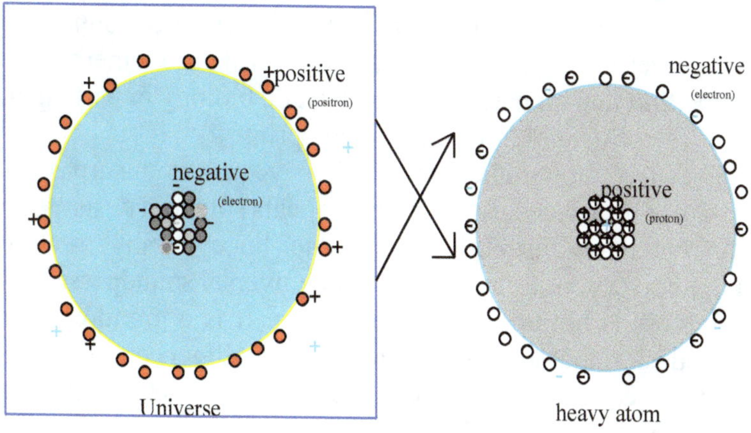

We should now be able to explain the universe's so-called *"missing mass"* by comparing its sketch with the sketch of a heavy atom.

Imagine the universal vacuum (left side above) containing a bunch of iron marbles piled onto each other. When you place a strong electromagnet around the vacuum, all the marbles will fly *toward the magnet* and stick to it. All we have done here is widen our perspective. Positronic energy defines the void boundary of the matter-universe, which has the physical property and dimension to act on us *like an oversized electromagnet, or better yet, "positron-magnet."*

Of course, there are differences: Universal bubbles are less solid than marbles, and the vacuum enables bubbles to grow larger. These bubbles grow into space; the positron-magnetic influence gets stronger and stronger. If we had the time to watch them *grow*, we would see more and more galaxies being pulled from within the bubble toward its edge to define it more clearly as time passes. This congregation of "electron-mass" outside of universal bubbles also determines their speed: the better defined a bubble is, the faster it should expand toward positronic energy.

Each region of space divides itself into these bubbles in the same way. A casual observer on Earth with adequate instruments can only look in wonder and amazement at the string of pearls in the depths of space. By visually connecting the pearls of light, a "honeycomb" structure emerges—a strong hint that all nature, in general, and biology in particular, adheres consistently to a cosmic blueprint.

The structural soundness of a honeycomb is unsurpassed. Perhaps this assures us that nature repeats splendid designs, be it on a small or large scale, and that the "honeycomb" structure of the universe is a sign of structural universal soundness. This soundness is not an accident of nature. It is a grand cosmic order design that guarantees universal Intelligence's survival. Cosmic equals positronic; universal equals electronic Intelligence.

Once again, since a picture explains a thousand words, let me save a few pages. Here is the principle idea of how a universal bubble may appear to have a "honeycomb" structure:

By visually connecting the observed galaxies and observing their distribution in a specified region of space, we see one structure that resembles a honeycomb:

FIG. HC–3:

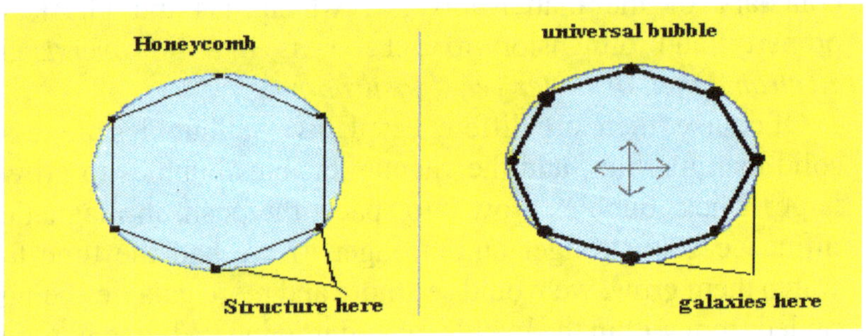

In 1988, astrophysicists discovered the "Great Wall," a cluster of galaxies at least 500 million lightyears across. The galaxies' redshift revealed a velocity of roughly 700 kilometers per second (437 mps), moved by the gravitational

(?) pull of some vast mass that scientists called the "Great Attractor."

The Great Attractor could be a tremendous force that no existing theory can explain. If gravitational pull were responsible for the fast motion of the Great Wall, we should be able to find this source of gravity with current instruments. We can't.

Mainstream scientists may need to be closer to the tree to see the forest. They need to specialize in small areas of their field and catch up on other connections. The pressure to conform and adhere to almost ancient "traditions" keeps their minds shut to bold new possibilities.

WHY NOT THINK BIGGER THAN ANYONE?

Last but not least, there are the dangers of political correctness, religious dogma, and simple ignorance.

Margaret J. Geller of the Harvard Smithsonian Center for Astrophysics, one of the universe-mapping scientists, was quoted in *Scientific American:* "The new surveys on the mapping of the universe are...impressive, but the state of our ignorance is equally impressive."

She is right. The history of ignorance regarding the universe is well established. Imagine that the Greek philosopher Leucippus already perceived a philosophy of "atomism" some 2500 years ago.

One of his first students was Lucretius, who was so impressed by the atomism theory that he said about 450 *before Christ:* "If the whole universe consisted only of atoms and voids, it was not infinitely complex but somehow intelligible, and there might be no limit to man's power...!"

Interestingly enough, the Greek philosopher Plato was about 22 years old then. When he later developed the groundwork for political societies, he gave rulers a blueprint for deceiving their citizens. It is no wonder that the ruling class considered the atomism theory dangerous from then on. They suppressed it with all available means for millennia to come. The punishment for believing in the existence of *atoms* in seventeenth-century France *was the death penalty.*

Is it a wonder that 21st-century science is intentionally or subconsciously shying away from "atomism" in general and a dual-versal model without the Big Bang in particular?

These new ideas give us the mandated and correct universal model. It would explain our true origin and purpose on this planet. Einstein sensed this honestly, but the scientific establishment quickly sidetracked him. The sudden emergence of elementary particle physics and the 1926 Copenhagen interpretation of Quantum physics gave rise to subjective mysticism, distracted from the atom, and confused everyone with irrelevant details.

. . .

Chapter Eight
Nuclear Fusion and Gravity:

After Albert Einstein's general theory of relativity, *nothing* meaningful emerged from universal physics because "quantum physics" started to confuse minds. Nobody could reconcile Einstein's theories with the "New Kid on the Block."

Einstein's gravity theory replaced Newtonian physics and was accepted; Scientists assumed that Earth-observed gravitational effects apply to the universe. Examine where gravity should come from when space existed before introducing matter.

What is the difference between the earlier described "energy spheres" in primeval galaxies and the end product: finally organized galaxies? It is

The Onset of Nuclear Fusion.

Nuclear fusion is the universal survival mechanism of matter that maintains a steady energy quality for bridging vast distances in existing space. Fusion is the practical lifeline of matter to survive its journey toward the force from which it originated and to which it will return.

Nuclear fusion started to emerge when the heavier elements, formed within a nucleon gas cloud, acted as condensing points around which lighter elements could form like the layers of an onion—the lightest elements forming the mantle of the star. This initial attraction resulted from the electromagnetic attraction of particles with differing charges.

When the cumulative electromagnetic *inward attraction* of an "energy sphere" became overwhelming, the electrons of the core atoms were slowly squeezed away and forced to build a ring layer around the nucleus. The pressure on the nucleus crammed the protons next to the protons to the point of touching each other until they began to *fuse*.

FIG.8—1

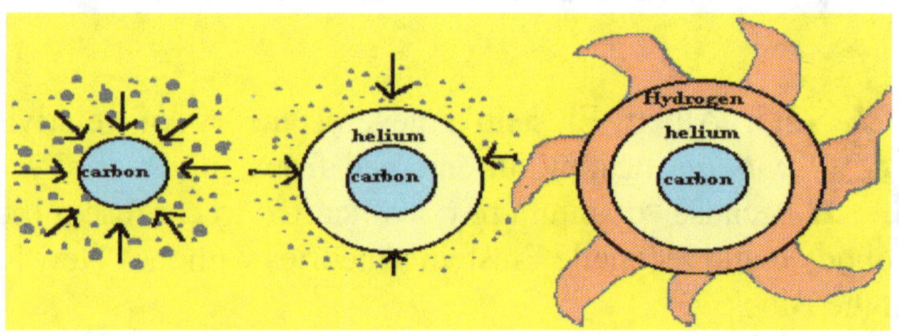

The electromagnetic quality of heavier elements acts as an attractor, around which the multi-layer of subsequently lighter elements can build up, the lightest outside. In the example above, Carbon is the heaviest element in *that specific nucleon cloud* and acts as an attractor to lighter elements, which start to build large layers around the carbon core.

When enough material accumulates, the cumulative pressure on the carbon core increases to a point where the electrons of the carbon atoms separate from the protons but can't escape. *They build an electron ring around the carbon nucleus.*

FIG.8—2

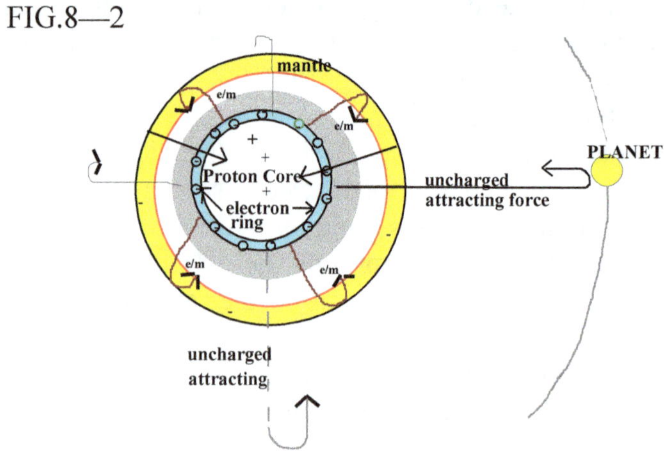

The cumulative inward pressure of the electromagnetic attraction of all layers toward the core *ignites the fusing of the core. A force emerges from the core that stabilizes the radius of the star and reaches out into space as an uncharged but attracting force:*

This force is what we call "Gravity."

. . .

A New Concept of Gravity:

It may be a revolution to view gravity as a *consequence* of nuclear fusion, flying in the face of part of the general theory of relativity. However, it is the only alternative *to* explain the disproportion of observed gravitational effects compared with the universe's total Mass.

This concept could prove that our current understanding of the force of gravity on a universal scale needs to be revised. It could also prove that the discussion of so-called "missing" matter in the universe has distracted us from finding an essential simplicity underlying its structure.

Although a different explanation for the origin of gravity does not alter the observed gravitational effects within our galaxy or solar system, it will make a big difference if we look at the total gravitational impact of the entire universe.

Suppose gravity is not a size-dependent constant universal force but a *fusion-quality-dependent* one. In that case, gravity can not be an absolute force, as is currently being claimed, but is relative itself, meaning that a galaxy with a billion stars of average-quality fusion may generate much less "gravitational effect" than, let's say, a galaxy with a billion stars of high-quality fusion.

In other words, a small, fresh fusion star could generate a hundred times or more gravitational attractive force than a vast, old fusion star.

Current measurements of gravitational tugs and pulls in the far reaches of space are mere approximations of the natural forces at work. Our instruments are insufficient to achieve an objective degree of precision and certainty.

If gravity turns out to be a "relative" force (as in fusion—quality dependent), each galaxy would have its strength of gravity, independent of its size. Intelligent beings of one galaxy could only enter the gravitational "halo" of another galaxy *if their values were approximate*. It would clarify whether other civilizations can physically visit our galaxy or have done so already, even if we neglect the time and distance factor.

Let me explain. Imagine another civilization living in a solar system in a different galaxy. Their planet is ten times larger than Earth but at the same distance as a sun of our size. According to our understanding, the gravity on that planet would be ten times stronger than it is here. Biological beings there would not grow as tall as we do here: the force of their gravity would stunt the growth of bones and tissue. They genuinely stay closer to their ground. Evolutionary development would have adapted their entire system to that specific level of gravity. If they were to visit us in our conditions, they would feel like astronauts on the moon. They could even fly by flapping their arms. A prolonged stay would undoubtedly make them sick. Vise versa, if we were to visit their home planet, a 160 lbs. man would weigh as much as 1600 bone-crushing pounds. Walking a distance of 100 feet amounted to torture, and only sophisticated technical support would make a prolonged stay feasible.

We may be unable to land or quickly depart softly from such planets, and life without a space suit would be impossible.

In both instances, we have an evolution of life and Intelligence. Both could communicate with each other effortlessly. Both could visit each other, but barely. Living together would be impossible.

Let's imagine their planet is the same size as Earth, but their sun is much younger than ours. As such, it would immerse the planet into a much stronger gravitational halo – up to ten times as strong as ours, with the same effects as detailed above. Such a planet would have to travel faster around its sun to overcome its stronger gravitational attraction. That, in turn, influences the perception of time as experienced psychologically: the faster you travel, the more time you can put underneath the belt, meaning you age slower, and so on.

The discourse gives a small idea of what else might be involved in looking at life in our universe. It is, however, different from the purpose of this book to go into more detail. Let's stick to gravity and start a revolution in thinking:

We know that the total gravitational effects of the universe are much stronger than the Mass indicates. The new definition of gravitational origin allows us to claim that most galaxies must consist of relatively fresh, high-quality fusion processes. It also enables us to directly connect the force of gravity to the existence and interaction of elementary particles in the interior of stars. *It may establish the embryo of a Theory of Quantum Gravity.*

Add to this scenario the positron-magnetic attraction of the surrounding more significant energy quality (the positronic universe), and we can account for the so-called "missing" Mass of the universe without ever needing to chase the ghost of unproved "dark matter," "cold/hot dark matter," or other brain teasers.

This new gravity origin theory also leads to another quasi-revolutionary statement: cosmic bodies *without* a nuclear-fusing core, such as Earth and other planets in our system, *cannot generate their gravity but **are gravitational "induction copies" of the sun's gravitational strength** in whose gravitational halo we are embedded.*

Each planet's gravity strength in our solar system becomes an issue only when immersed within the gravitational halo of a nuclear-fusing star relative to its size. We determined that Earth's size is our value of one, with other planets compared to that according to their *relative* size to us.

What is more? This gravitational induction hypothesis is the point on top of the "i" for physicists. It would *add another consistency in the link of "inductions" already known and accepted: electromagnetic induction, heat induction, light induction, and now gravitational induction.*

· · ·

Chapter Nine
<u>Star Explosions and Supernovas:</u>

Gravity and its dependence on the quality of a star's fusion mandates two things: if the nuclear fusion of a star deteriorates, the star's gravitational strength deteriorates. Consequently, the orbital distance of planets around any sun will get longer, and more time is needed to compensate for the longer distance the Earth has to travel. At any further weakening of the gravitational strength, the planetary system may get disorganized, disintegrate, or drift into a different area within the galaxy. The discovery of more planetary systems may provide more explicit evidence of this postulate.

When a star's gravity (fusion quality) slows down, the outer layer either drifts away, at below 1.44 times the size of our sun (Chandrasekhar limit) or will be explosively "attracted away" from the nucleus into space *by the all-attracting force of the oppositely charged positronic universe of far away.*

The star's mantle can not cling in a balanced way around the nucleus, and everything down to, sometimes including the "electron ring," is "attracted away" from the core: The star "explodes." We could also coin a new word: *"attractplosion."*

The remains of this stellar explosion could be a quasar: a pure proton core with positive electric charge spinning under the long-range influence of positronic matter. If electrons managed to cling to the nucleus, it could become a "black hole-like entity."

Why?

When incoming photons (from another light source) collide with an electron and push it closer to the nucleus, the photon is absorbed, giving the impression that the light disappeared. When light continuously disappears in space, the region appears black (devoid of light). Given also the enormous density of this universal oddity and its overwhelming positive charge, it is possible that nearby "electronic matter" is gulped up the same way as light. It disappears behind a "cloak" whose core is impenetrable to electronic instrument detection.

Let's return to the point of explosion.

The mantle of the star disintegrates into elementary matter particles. The particles retain their electric charge. The resulting attraction to particles with an opposite charge mimics the same process as initially in the Big Conversion explosion. With time, distance, and exposure to space, all released particles of a supernova will eventually reassemble themselves into new energy spheres that again start nuclear fusion. As such, each supernova is a "mini"- Big Conversion explosion.

This continuous process of energy maintenance fortifies weak energies with fresh energies and prevents ordinary matter within the universe from cooling down too far for life and Intelligence to exist.

Not only do stars or planetary systems collide and explode to re-group vital energies. Whole galaxies can crash into each other, giving the weaker partner a second chance for life.

Our solar system, if not the entire Milky Way galaxy, is such a second generation of energy form, arisen like the phoenix from the ashes of the energies of a colliding galaxy some 4.5 billion years ago.

. . .

<u>Gravitational "Halo" and Galactic Collisions:</u>

FIG.9–2, below shows two equal size galaxies with differing gravitational halos:

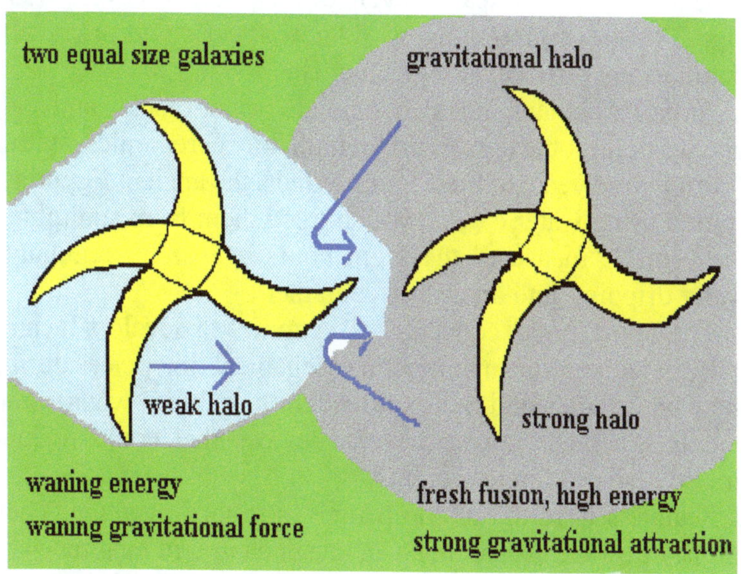

The weak galaxy is attracted by and toward the strong galaxy and fortified with new energies, requiring the surrender of its independence. As a biological occupant of either of these galaxies, I would arrange a temporary "leave of absence" and come to our neck of the woods. On the universal scale of things, however, temporary biological life is destroyed under these circumstances. It does *not imply that the universal consciousness of these biological life forms is destroyed or "dead." It survives, as we will see later.*

· · ·

<u>Time, Speed, Distance, and Intelligence:</u>

When explaining time, we must distinguish between our

somewhat arbitrary understanding of planetary (Earth) time and universal time.

Early humans only lived by seasons derived from observations of the sun's and Earth's relative motion. A repetition of a four-season cycle or a two-season cycle was a signal to start counting anew. A "new" year could practically begin at any position along that cycle.

Initially, humans invented sundials, water clocks, and sand glasses to limit the duration of certain holy ceremonies. Within five hundred years of their first sundials, the ancient Egyptians started using water clocks, which freed them from sunlight as they could "measure" the night hours by strictly regulating water dripping into an inflow container.

The "equal hour" concept took centuries to develop. It came only about with the invention of mechanical timepieces during the Middle Ages and their subsequent portability when smart people invented "spring-driven" devices that could replace gravity-driven counterweights.

Seafarers were the first to benefit from this invention. Their journey over the oceans along an imagined grid of longitude and latitude established a distance covered divided by the time it took, which anyone with a similar device could repeat independently.

The position of any ship could now be measured by the ship's crew precisely, as well as being repeated by the crew of another vessel at any time – but only by Wolfgang Pauli's 1925 formulated "exclusion principle": "two similar particles can not exist in the same state; they cannot have both the same position and the same speed within the limits set by W. Heisenberg's 1926 formulated "uncertainty principle'. Applied to the seafarers, two ships on an ocean can not have the same position and velocity *on one set of coordinates*. The more accurately we know the position of one ship (longitude/latitude), the less likely it will be that another ship could be at the same spot simultaneously (unless you stack them upon each other). As such, time, speed, and position make for an intrinsic universal value of every matter particle within our universe. They are a prerequisite for their existence in the real world.

The Earth-evolved "natural" concept of time is only a far-related cousin of universal time. Only understanding universal time can explain the "flow" of time as we experience it psychologically.

. . .

Chapter Ten
Universal Time:

Universal time comes into existence with the formation of matter particles in three-dimensional space. Here, we assume the possibility of almost linear expansion of matter particles into existing space. "Almost linear" because the energy rotation before the Big Conversion explosion gave all particles an outward-spiraling forward motion. See Fig. 10–1 below:

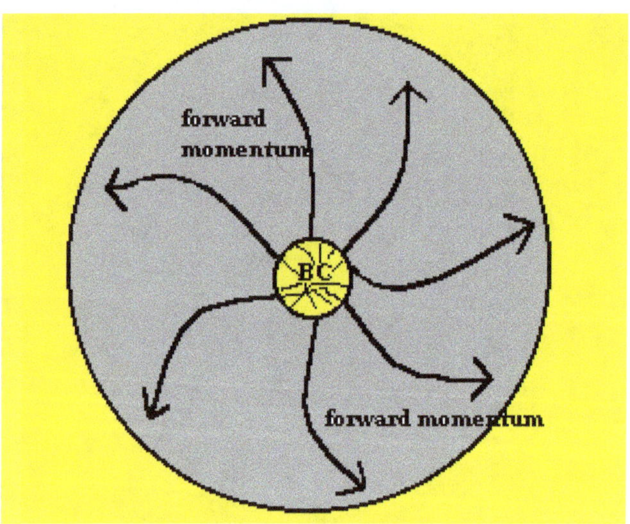

Each position of a matter particle along its flight path determines the coordinates of three-dimensional space: longitude, latitude, and depth. This position changes continuously due to the particle's forward motion. (particle stands for particle, planet, star, galaxy, or galactic cluster.)

Each tiny change of coordinates is called an "event."

The particle's path can generate coordinates of already experienced (known) positions, which I call "activated events," and coordinates that the particle will eventually reach but are yet unknown. These yet-to-come positions are named "yet-to-be-activated-events." As such, each particle "creates" its own time expressed as the ratio of already activated events to yet-to-be-activated events. Its proper position (actual) is called the "Now" position.

The "Now" position is fleeting, lasting less than a nanosecond. (Also called 'present')

. . .

Event Conversion

Each activated event has its specific position in space, and this is one reason why we can't see a broken cup jump back onto the table to reassemble itself in real-time, as we could see in a film of that event running backward. The time it took the cup to fall *from the table to the floor* coincides with a continuous forward motion (of the galactic system) that ties the point of breakage of the cup *to a different position in space* from which we have distanced ourselves already (literally). This *point of no return* is called the past (passed) event. We have experienced it and can remember it.

Positions not yet reached in space are not yet connected or "activated" by motion - events we call *the future.*

This sterile explanation of the physics of 'event-conversion' or the 'flow' of time in one direction does not justify our perception of *'duration'* within that flow.

During unexpected, sudden, or extreme situations, such as serious accidents or "near-death experiences," time slows down to a crawl like molasses on the North Pole. Each second seems to last a lifetime, each minute—an eternity.

This perception results from increased consciousness in our brain, which now works almost at the speed of light to find a way to correct the situation. A fast and intense brain seems to slow down all occurrences around you.

This observation fully complies with the laws of physics and the problem of 'time dilation,' as Einstein had already theorized. According to that theory, time slows down for a fast-moving object, as measured by an observer, which is considered stationary or relatively slower, also called the "twin paradox" in which one member of a set of twins traveling at or near the speed of light would return home from a distant star and find that he has stayed much younger in comparison to his brother who had stayed at home.

The existence of this time dilation and the fact that we can measure it have yet to lead us to the discovery of its origin. An explanation of the Hubble Constant could change if we focus on its most important single component: *speed.*

Edwin Hubble established in 1929 that a galaxy's spectral redshift is not random but directly proportional to its distance *from us*. The farther away a galaxy is from us, the faster it moves. This discovery led to the conclusion of an expanding universe, meaning that *a continuous, equal-segment event conversion is physically impossible in our universe.* In an expanding universe, it takes less and less time for a moving system (and everything is in motion all the time) to convert a fixed amount of events. If galaxies farther out in space are moving faster away from us, we could say that their relative position is *more advanced* and that *a ratio between advanced position and speed exists*, meaning that the speed of our galactic system (the Milky Way) must steadily *accelerate as we speak* to achieve the higher speed of a positionally more advanced system *once we reach that position in the future.*

This continuous acceleration of our galactic system gives *each generation of biological life* on Earth a different perception of the "flow of time."

Or let me explain it this way:

Our galaxy is currently moving at a speed of approximately 42,000 miles per hour into existing space. It may have moved with only about 30,000 mph a few million years ago. "Continuous acceleration" means that things are speeding up and that our system will move in the future with 100,000 mph, then 200,000 mph, a.s.o. A fixed distance of, let's say, one million miles is bridged ever faster. What takes us 24 hours now to clock one million miles would take us in the future only ten or twelve hours.

The *birth* of a living organism on that accelerating system always coincides with the *slowest and actual speed of that system* in that organism's lifetime. Infants don't even have a concept of time. It is a strange new "thing" that they slowly use by observing the actions of others like the mother returning with food at specific intervals. For older children, time *creeps in.* Each hour is 'timeless.' Every twenty-four hours is a whole life lived. A week is an eternity.

Yet, for the adult parent who watches this child grow up, time seems to fly much faster. It does. The adult was born at a time of much slower 'event conversions'—at a position of the planet of years ago. His "now" position is much more *advanced* relative to that of the child or to his own time of birth on that accelerating system.

Suppose we now put a grandparent side by side with the just-born child. In that case, we will have two distinctly different entities: one who has experienced a long trail of ever-faster-occurring event conversions, one who is just starting to experience event activation, and one who has many yet-to-be-activated events to feel and experience.

The difference between the two is universally significant since we used to call it the "generation gap." It explains *why* a genuine bridging of that gap in terms of identical feelings and accurate and equivalent understanding of things and one another can only be approximated.

This phenomenon also explains why a young child initially needs so much time to get used to our mind conditioning and sense of humor about the reality of our world. As a fresh "creation" of the universe, it instinctively knows the true nature of the universe and resists, in its very own way, the inculcation of values that are known to be wrong. It takes about five years to "unlearn" this resistance and prepare it to accept the magical new world with its restrictive rules, grasp a sense of the momentum of motion, and feel and experience the unknown concept of the progression of time *that the child cannot learn from reading a book.*

Our task as educators of children and adults should be to emphasize one universal truth: It is only possible in our universe to gain higher insight and knowledge by also being aware of the positional advance of the galactic or universal system. Combining both transforms personal awareness into universal consciousness, which comprises the highest state of biological Intelligence.

Time segments, or event-conversions, can now be expressed more accurately for all events between the completion of the BC (Big Conversion) and the onset of BC* (Big Crunch).

The horizontal middle line in Fig.10–3 not only represents the objective universal time from a slow start (on the left; the past) to a fast staccato in the future (to the right and expressed in the different spacing of the vertical lines).

The entire principle also reflects our biological passage. The perception of time from the point of gaining independence (birth/infancy = slow) is also expressed from left to right by the vertical bars' spacing. The older we get, the closer together the vertical bars. Spacing equals perception of passage or flow of time:

Fig. 10–3:

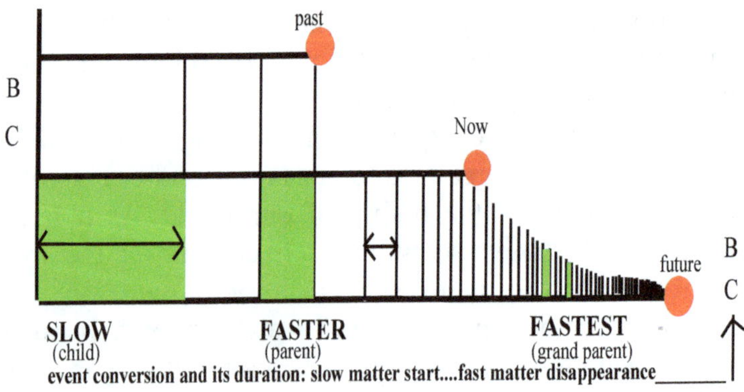

"Passage of time" and the perception of the speed of time flow indicate *advancements* in space and individual Intelligence. They are firmly intertwined. Intelligence is an expression of the experienced number of events—conversions at ever-faster speed: the longer you live, the more you know of this world. Think about it.

The following sketch shows "us" and our "now" position in comparison to 'less advanced' and 'more advanced' galactic systems.

The less advanced galaxies are moving at a relative slower speed, and the more advanced with a relative higher speed when compared to our position:

FIG.10–4:

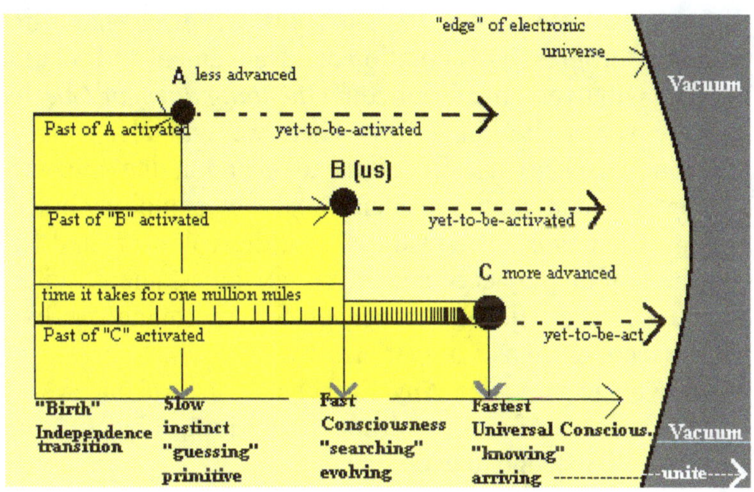

This diagram shows that the closer we come, or would come to the 'edge' of the universal bubble, the more advanced we will be. Personally, there is nothing to worry about the "edge" as it has a head start over us of billions of years. What remains to deal with is our accelerating motion.

As a matter of observation, we know that the more advanced or older we are, the more knowledge we accumulate. Knowledge accumulation is proportional to the speed of event conversion: the faster the galactic system moves, the faster we accumulate new knowledge, and so it derails the adage, "You can't teach an old dog new tricks." Whether you like it or not, you can't stop learning new things, young or old. Although there are differences in the speed of knowledge accumulation between young and old in small specific areas, such as languages and short-term memory, the bottom line remains the same: what the young accomplish with enthusiasm, the older compensate for with experience.

Humanity is currently reaching a threshold of higher consciousness and begins seeing the ever faster changes as they enter daily lives. People who can discriminate and adapt quickly propel themselves to the forefront of human evolution, education, and the acquisition of *universal knowledge*, which can not be interrupted by the tools of electronic mind pollution. Progress as a condition of advancement and position in space will overcome the ongoing *dumbing down of America and the world.* Within a few years, we will also reach a position in space that forces us to abandon the medieval concept of accumulating financial wealth by exploiting others' biological energies. The result will be the voluntary and free exchange of already accepted universal values, and acquiring objective and applicable *knowledge* for improving our lives will be recognized as a higher priority.

Knowledge is power. Any further advance of our system into the reaches of space will inevitably bring objective and universal knowledge to everyone.

Positronic energy is acting on us as we speak to realize positional advance and practically instill in us a purpose that equals or exceeds any power we humans may ascribe to a force called "god" or "government."

The positronic energy is infinite in size. It uses part of its energy to convert itself temporarily to matter. It does so to maintain its energies indefinitely. It has given matter-energy an inherent limitation within the vacuum: the limit of the speed of light. This speed limit allows the formation of transitional biological life, which can experience the joys, sorrows, and objective riches of a larger universe that always exists in infinite generations.

The moment in which we understand that we are the living manifestation of the physically existing *positronic universe*, we can discard all notions of limited resources. We then can become *universal* beings. Universal beings don't exploit the energies of others for their selfish interest. They don't steal, rape, maim, or kill their fellows, friends, and neighbors or destroy the rest of nature. Most importantly, this feeling of universality will creep up on us overnight in a velvet-revolution. At that point, we will ignore the mainstream media, social media, and our governments. All this without bloodshed or force.

By doing so, we could immediately free enormous scientific, financial, and creative resources worldwide. Adopting the new cosmology introduced here comprises minimal and meaningful knowledge from which we could explore the powers of the positronic universe.

By extending our horizons to include positronic-electronic dualism, we could replace dogmatic mysticism with a meaningful philosophy based on up-to-date knowledge of universal reality.

Progress has never been easy or automatic in our stagnant, dogmatic world, maintained by force. We don't need more force but more solutions.

Forbidden Cosmology is a valuable foundation of essential solutions. It is an umbrella philosophy that accommodates all mystic and religious ideas as sub-aspects.

The next chapter lays the groundwork for this umbrella philosophy by showing the connection of material physics to the beliefs of our ancestors. There are remarkable similarities. We owe it to ourselves to objectively explore these connections desperately needed to secure our future.

• • •

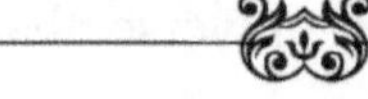

Chapter Eleven
The equation Einstein missed:
(Highly speculative)

$$EE' = M\,c^* \quad (* = n\,/\,\text{infinite})$$

(pure energy equals Mass at infinite speed)

For the first time in history, a scientific model of the universe derives from the laws of physics and observational evidence that suggests a purer form of energy from which we originated.

This dualism—image is part of our brains, mainly consisting of hydrogen, oxygen, and carbon atoms. Their very structure, the protective electron envelope, gives us an independent power of recognition. This cognitive power develops slowly and depends on our positional advance in space.

The early man sensed this enormously attractive force but needed more scientific knowledge and instruments to interpret it correctly. With the tools at hand—almost nothing—they had to explain natural phenomena in a context out of which organized religion eventually developed.

"Forbidden Cosmology" explains with consistency the theological components of creation: energy, light, a pre-existing form of Intelligence and universal consciousness, and the conversion process from that energy to our matter.

A. Einstein developed his theories of relativity between 1905 and 1915, but he didn't consider the topic of Antimatter or positronic energy then. We confirmed Antimatter in the early thirties. Even then, theorists and particle physicists were slow to change their minds about the discovery's strange qualities, and many incorrect "symmetry theories" emerged. All theories to date assume the exclusive existence of the matter universe and nothing else.

Quantum physics started to develop the most complicated *theories* of Antimatter's existence and how it emerged in the energies of the early universe. We know that particles can turn into antiparticles at high energies and vice versa, and this process took place right after the Big Bang explosion. We thought that the initial rapid cooling of the universe led to the destruction of antiquarks by quarks and that a slight excess of quarks remained from which the matter in the current universe arose.

The CCEEC (Complete Cosmic Energy Exchange Concept) confirms the quantum mechanical assertion that antiparticles turn into particles at very high energies. *This assertion implies that changing a particle's electric charge leads to a quality change. (Antimatter becomes matter, and matter can become Antimatter at very high temperatures.)*

This thesis reversibly shifts the process from events *after* the Big Bang to events *before the Big Bang*. More precisely, the Big Bang *results from the conversion process from Antimatter to matter particles*. This conversion left a clean slate for matter to develop stars and galaxies exclusively without interference in an existing vacuum.

Scientific dogma's resistance to adopting this elegant solution is regrettable and incomprehensible. The results shown are *consistent with already established ideas*. The general public would not question science's objectivity if it were to redraw its battle line along some newly discovered consistency.

The impact on religion could be more significant. The mythical belief in a personal God who created the universe and humanity would not stand the test of time. It replaces a physical conversion process in which existing positronic energy literally "created" electronic matter.

However, the beauty of the positronic model is that *the "power of God" can be taken out of the Big Bang and transferred back into space to include the positronic universe.*

Positronic energy is all-powerful, always existing, and knows all of its creations, just as believers in God and creation maintain giving a "God" to atheists as well. They could see in it a form of existence that adheres to the laws of physics but also inherited Intelligence from which an open mind can draw strength when needed.

Subconsciously, we have long confirmed the existence of the positronic universe. We connect with it via our thought process in extreme situations. Enough anecdotal evidence has been collected and published about the occurrence of "Near-death-experiences," or "NDEs." In extreme conditions, such as accidents or complex surgery, the biological death of an individual may occur for short periods. When medical personnel try to revive the patient, he sees himself leave his body and float away. A black tunnel opens filled with a soothing, peaceful, white light that attracts.

Depending on the expectation, education, or religious upbringing of the person entering the tunnel, the light contains all those things that were dear to the person during their lifetime. A devoted Christian might see Jesus Christ or angels. An environmentalist might see lush forests and crystal clear water. A child might see their favorite toys, angels, pets, or grandparents.

The universe's infinite size can most likely produce everything imaginable with a human mind and more, without a limit, in total objectivity, and without blame or judgment—in short, what we all envision "heaven" to be. This energy is love without authority. If you are a sinner or saint, you are welcome because this energy is responsible for your existence.

It is the energy from which we originated.

Near-death experiences often end interestingly: the peaceful, objective energy sends some people back into their biological "shell," their body, to resume a yet-unfinished task.

FIG.3–1 of this theorem shows the exact mechanism of the tunnel between our world and positronic energy: a particle-free vacuum into which the matter universe expands, and the void acts as an insulator between the two energies. No biological form can cross this void; only our consciousness can connect.

You can only connect to something that already exists in reality and as a reality. It indicates that universal consciousness exists as a field – a force that permeates the space around us.

We know this is possible because we have visions and dreams that appear out of nowhere. Sometimes, we get *a brilliant idea* without consciously trying to get it. "The bulb is lit!"

Who or what could do something like it? Does it come from this elusive, ethereal force we have been trying to find for two thousand years to no avail?

Ask me, and I will say *it is the positronic universe.*

It is hard to imagine its energies. No words in any language could fully describe it. Imagine an energy much more efficient than nuclear fusion. Imagine temperatures so high that not even atoms can form. A structure in which elementary antiparticles exist in such a high *vibratory* state that words like "speed, time, or laws of physics" have no meaning whatsoever.

This energy has utilized electronic matter and its Intelligence from millions and millions of other universal vacuums, which are not connected. Stop and think about this for a while.

The very construction of matter with the negatively charged electron on the outside (of atoms) enables it to attract its original counterpart, the Positron. We call their meeting a "thought." Therefore, only a *conscious mind* can instantly connect to that energy. This energy is around us and above us. We should not leave anything to chance; the location of our brain, at the highest point of the body, is rightly designed to facilitate this direct connection by facing each other.

Positronic energy is the epitome of purity, unlike anything achievable within the electronic universe. It is pure elementary antiparticle energy or **E'.**

As explained, pure energy, E' contains the extract of electronic matter, **E**, since it has utilized the matter from countless electronic universal vacuums. The full name for pure energy should be E + E', which becomes **EE'.**

Infinite size and heat imply an infinite *chance grouping of elementary antiparticles in a plasma-like existence.* They practically waft, vibrate, and pass each other, and by their sheer numbers, they can create gravitational or mass-like effects. These effects then pass through space as a <u>field force</u> (permeate it at all times). This field force has a measurable impact on all matter in the electronic universe.

Look again at Fig.3C in this manual. Our universe becomes a dot in a "landscape" whose existence we deny to date. It should be evident to any viewer and reader that a field force can easily permeate a dot like this. This larger perspective does not alter our developed sense of time, distance, and the possibility of a long biological life.

By allowing an honest exploration of the actual structure of the dualverse, we could prove scientifically that positronic energy has the needed dimensions to generate mass-like effects that can permeate space. This quasi gives it the quality of MASS, **M.**

The vibration of elementary antiparticles exceeds any motion achievable by atomic (electronic) matter and is, as such, not limited to the physical limits of matter. The speed of these elementary antiparticles in an environment other than a vacuum in which *we* exist is practically infinite. I call this *c/infinite*, or limitless speed (instantaneous). It denotes an almost instantaneous communication effect in all directions, thus leading to another quasi-revolutionary conclusion that is hard to fathom in our state of intellectual development: The positronic *energy* (EE'), which equals Mass (M) at limitless speed (C*), *functions like a brain and <u>IS</u> a brain. As a brain, it equals thoughts and Intelligence.*

The brain, thoughts, and Intelligence of its conversion product—matter and biological life—can still perceive its existence and connect with it. Contrary to anything you may have heard or learned, *it is neither mind over matter nor matter over mind.*

In the loosest translation of cosmic energy, mind equals matter, and matter equals mind.

By finding the missing part of the physical universe, we have also found a fitting extension to Einstein's equation E=Mc2:

The symbolic equation for the positronic matter would be:

$$EE=Mc* \quad (*=n, infinite)$$

Pure energy equals Mass at infinite speed (instantly everywhere)

EE": pure, positronic energy containing the energy of reclaimed electronic matter;

M: Mass-like effects of positronic energy which permeate our vacuum as an attracting force;

C*: the unmeasurable agitation of positronic energy which communicates instantly in all directions

"positronic": elementary antiparticles with an opposite electric charge to that of electronic matter, not suspended in a vacuum, and not bound by the limitations of electronic matter (speed, motion, density, gravity, etc.)

· · ·

<u>Cosmic Microwave Background Radiation:</u>

In 1965, the American scientists Arno Penzias and Robert Wilson of Bell Telephone Laboratory in New Jersey detected faint microwave radiation that permeated space evenly from all sides. They concluded that light from the Big Bang had traveled so long to reach us that it now appeared to us as microwave radiation (wavelengths of less than one meter). They were awarded the Nobel Prize in Physics in 1979 for their discovery.

The discovery of this background radiation led to great enthusiasm among cosmologists and astrophysicists, as it seemed to confirm the Big Bang *theory* itself. Many years and numerous satellite launches later, the mystery of cosmic microwave radiation became ever more profound.

Penzias/Wilson assumed that any radiation *in* our universe must also result *from that* universe. We set out to prove it. In 1989, the US launched the American Cosmic Background Explorer satellite to bring us more details about this radiation. The plan was to look for temperature variations that accounted for the later formation of stars, galaxies, and galactic clusters.

To everyone's disappointment, the satellite did produce evidence of a near-gloss-slick background for the universe. We discovered no noticeable or out-of-the-ordinary detectable fluctuations that would have explained the early structure of the universe.

In June 1990, Germany launched a Roentgen satellite (Rosat). It found, within a few months, more than 80,000 new microwave and X-ray sources, including unexpectedly ancient quasar clusters in the early universe. Large clusters in the early universe are a problem in explaining the aftermath of the Big Bang. Scientists believed that galaxies and galactic clusters drew slowly together using gravitational force. This process should have taken billions of years. If we suddenly find a large structure in the very early universe – about a billion years after the so-called Big Bang, then the structure must have come about by other means (see earlier chapter on energy conversion).

More recent inquiries into the microwave background radiation suggested that it may have nothing to do with the Big Bang. The question of its origin then gets even more complicated. Could it be that it doesn't originate from our universe?

How could we explain the redshift of the cosmic microwave background radiation, meaning it is rapidly receding from us, with my conclusion that it comes from an external force *toward us*, which would mandate a "blue-shift"?

The dualversal concept may *shine* some light on it:

FIG.11–1

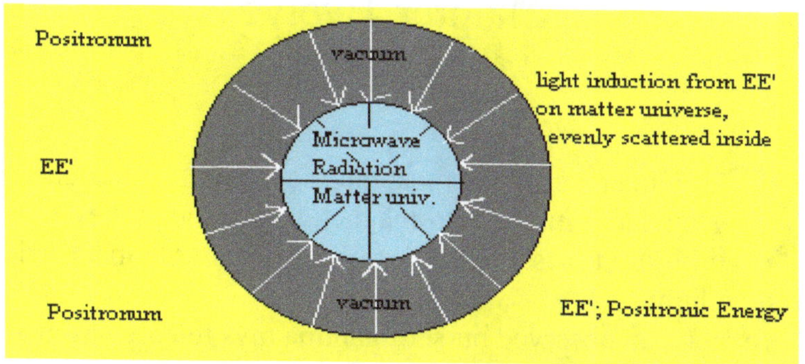

The "light induction" from the positronic energy on the advancing edge of the matter universe forms the near-gloss slick sheen that helps to define that 'edge,' and we detect it as gamma, X-rays, and microwaves.

Fig.11–1 shows why microwave radiation smoothly permeates space from all sides. It can exist regardless of the size of galaxies or clusters and their age.

By focusing our instruments from within the expanding universe toward the edge (relative to our position), the microwave background radiation is receding and appears redshifted.

. . .

Chapter Twelve
Gamma Ray Bursts:

Gamma rays are the most intense form of matter-energy on the entire electromagnetic spectrum. They are a billion times more potent than the particle that carries visible light.

In 1979, an observed burst of gamma rays released as much energy in one-tenth of a second as our sun will release in every form over the next one thousand years. Wow! The short duration of these bursts makes them hard to study, but they occur almost daily.

Gamma-ray bursts have been called the "astrophysics mystery of the decade." The puzzle is that scientists are still determining where the gamma rays are since they seem to grow in open space, unconnected to any nearby or known objects.

At a meeting of the American Astrophysical Society in January of 1994, scientists presented evidence that some gamma-ray bursts occur throughout the universe and as far as five to ten billion light years away. We can only solve the mystery once we acknowledge the possible existence of a tangible, physical universe as described in this hypothesis. By doing so, we would extend the scope of the electromagnetic spectrum by 100 percent, doubling it to accommodate the frequencies higher than gamma rays.

The phenomenon of unexplained gamma-ray bursts, coming evenly from all sides and regions without galactic structure, has the same underlying principle as that explained in the origin of the cosmic microwave background radiation. While traveling long distances, pure energy interacts with the advancing 'edge' of the matter universe and scatters into the highest energy/shortest wavelengths, which we can detect. Deflection or absorption by the advancement of galactic systems creates turbulence. Since intergalactic space is not entirely free of single particles (a host of gas clouds swirl all over the place), positronic energy plows into these clouds (invisible to us) and "eats them up in a flash." Depending on the volume of particles in a region of space, the bursts can be catastrophic, firm, or feeble.

. . .

<u>The Phenomenon of Consciousness</u>

Neuroscientists, philosophers, and physicists are intensely discussing the phenomenon of consciousness. The general consensus is that the human brain is the most highly evolved biological matter within an exclusively existing single universe. For this reason, the physical seat of consciousness could eventually be located. Different targets have been proposed for further investigations.

Roger Penrose is investigating mysterious quantum mechanical processes within the brain's protein tunnels, called microtubules. Nobel laureate Francis Crick wants to study the neuron connections responsible for visual awareness. Andrew T. Weil, an Arizona physician, proposes that a complete theory of the mind should include an explanation of why South American Indians experience identical hallucinations after ingesting psychedelic drugs.

Only this is certain: the brain generates effects that make us aware of our existence, feelings, and thoughts. However, searching for a physical location of consciousness is a perilous and questionable endeavor. If any specific brain region could be declared the "winner" in the generation of consciousness, it would become a target of extinction.

Government-financed scientists would not hesitate to disrupt these brain regions in people declared enemies to the "authority of the state." A crop of zombies who function as workers and vote-givers in elections but would otherwise be devoid of their sense of specialty and the actual reality of their lot in life could be created. It is no secret that military scientists

69

are at the forefront of these studies to use them, you guessed it, *against* our own population.

Fortunately, no matter what they do, they will fail.

The recognition of one's specialty and personal awareness does not depend on a specific site in the brain. Instead, it is an effect *of the total sum of the brain, which acts as a receiver of externally existing forms of universal consciousness.* The discussed dualversal model predicts, and history has confirmed, that sudden leaps of consciousness occur in the entire population, regardless of which authoritarian regime or educational system is in place. This observation would then tie consciousness to the forward motion of galaxies and the universe's expansion into new regions of space. It is a process in which the brain can connect to other forms of consciousness at *predestined positions in that space.*

The universal vacuum is permeated with the positron-magnetic field forces of the surrounding universe. The strength of this field force correlates to its distance to positronic, intelligent energy. The closer we come to it, the higher the quality of our consciousness.

FIG. C–1

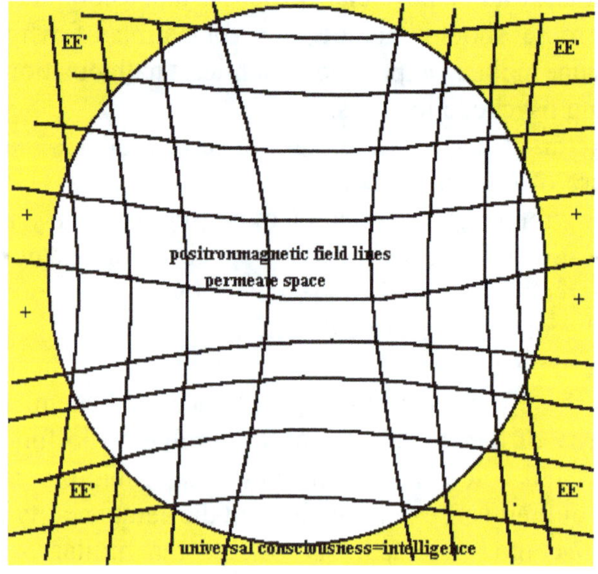

Let's again examine the physics of the model, which parallels the evolutionary biological progress observed.

The previously discussed "Penning-Trap" of Hans Dehmelt used an electromagnetic field to tightly hold an electron at the dead center of the trap. By retaining the charge of the "electron" (the electronic universe) and by reversing the outer charge, which practically floods the whole field with positron-magnetic lines, something very predictable will occur: The entire electronic universe, which has a net-electric-*negative* charge, will be attracted toward the wall of the universal sphere, which has a net-electric-*positive* charge. Can we make it any simpler?

This positive/negative or positronic/electronic attraction accomplishes two functions. First, stoppable expansion of our universe, and second, motion brings us into new regions of space. These new space regions are permeated with ever stronger universal consciousness emanating from the originating energy.

The proof comes from our development.

Human evolution shows a clear and linear progression from natural instinct some 10,000 years ago to the development of the bicameral mind some 6000 years ago. The onset of human consciousness occurred around 3000 years ago.

These were times of enormously powerful changes.

When society became increasingly complex three thousand years ago, the prevailing bicameral mind—meaning the separate functioning of the two halves of the brain—could not handle the pressure and would have led to society's premature destruction. As if to rescue mankind, (human) consciousness sprang to mind.

The managers of these societies immediately recognized it as a powerful tool. The formulation of Western religions took place as a more subconscious definition of these universal forces. Unfortunately, the newly developed religious ideas were immediately used to institutionalize the "divine" mandate of authority with its 'right' to impose restrictions on the general public.

This justification, in turn, was used to lead the public astray and distract it from the ongoing unscrupulous confiscation of individual and collective wealth.

The slight advantage of knowing more about the universal working mechanism could be maintained by keeping most of the population as busy, distracted, and uneducated as possible. A wall of illusion went up with the development of "moral" codes that denied anybody, except "God-chosen" rulers, the chance for a joyous or dignified life. While rulers took whatever they needed, from material riches to sexual needs, their subordinates were "rewarded" with nothing but empty promises, and even these promises were pushed into an unspecified "afterlife" with its (artificial) power of judgment and condemnation.

Come to think of it, isn't it how we are still expected to live and behave today?

The time of illusions, deceit, and distractions is now ending. One more time, the upholders of "tradition," illusion, tricks, and deceit are calling to arms to keep the world enslaved. Under the disguise of "spiritual revival," a worldwide religious recruiting drive is underway to corral the dummies who were designed to obey and who would not hesitate to jump off a cliff or stab their neighbor if ordered to do so.

Fortunately, this "back to the past" movement is counteracted by a new quality of consciousness emerging worldwide. This new quality is called *Universal Consciousness,* which is higher than human consciousness and has prevailed since the time of Greek philosophers.

By its very design and *cosmic origin,* it can *not* be misused as a political or commercial tool of oppression and exploitation. To date, an explanation of this phenomenon needs to be improved in a language anyone could understand.

The dualversal cosmology is a simple, accurate, scientific, and easy-to-understand model that could become or act as the catalyst for essential worldwide change. Many will embrace it as a tool to implement peaceful and positive social changes.

It will help us overcome wrongful theories of creation and

evolution and lead to the end of all practices of mind conditioning and suppression.

What is the model of consciousness that can bring these changes?

Here it is:

The objective *and existing layers of consciousness*

FIG. C–2:

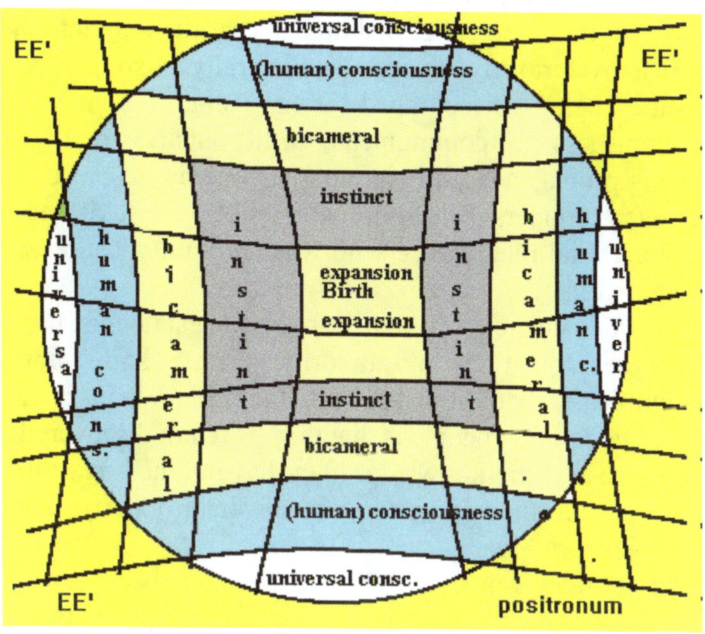

Please take the time to study the sketch above. It is a perfect example of consistency in connecting seemingly unrelated topics—biological consciousness and the universe's expansion—with the stroke of a pen. Not only are the two categories mentioned entirely interchangeable, but other facets of life fit right into the same sketch. All we need to do is replace the "label." This amounts to the ultimate test of "Truth" when *honesty* is applied to it.

The denial of this objective system by mainstream

scientists, sociologists, politicians, and religious leaders is a powerful form of mind conditioning meant to uphold the division of mankind and the mandate of the "stratification" of society. If not changed soon, it will lead to an implosion of all societies because all forces are set against each other in vain attempts to fight for their "cause." Innumerable extremist groups will emerge who have no reasonable awareness of the manipulation of their consciousness by those who use them as a tool of terror and intimidation.

By enforcing the belief that humans are the only intelligent life in a mysteriously "created," unilaterally existing universe, we are being discouraged from using the power of consciousness to communicate with outside intelligence. Today's propaganda aims to label anyone as far-out if he or she even ponders or proposes publicly the existence of extraterrestrial intelligence with which communication *might be possible.*

I am *not* a conspiracy theorist, but I must state this: those in the highest positions of authority already know how to communicate with outside intelligence. What is most frustrating to these people is that they have not discovered *any way* to misuse the knowledge they have gained against you and me. What they are getting is universal honesty, making their faces turn blue.

Our politicians-magicians have created a monster state with trillions of dollars in debt just to pay for weaponry that was never intended to be used against an outside enemy. The signs are apparent. Although the Cold War ended years ago, the military budget is going *up, paid for by starving American children and a semi-literate, working zombie-class.* With the onset of universal consciousness, our politicians have all reason to fear the future. They know their game is ending when each of us discovers their knowledge and insists on being "plugged in" again.

Under the danger of repeating myself, I state again that this fear inside our and that of Western governments borders now on paranoia and clearly led to the decision to terminate all "SETI"–projects, to intentionally launch a blind Hubble Space Telescope, and to end all deep space exploration. They fear that the information-speed and photo quality of these technological marvels will reveal the existence of extraterrestrial intelligence, and the bureaucracy is not prepared to stamp a "Top Secret" on all incoming electronic data. It is much easier to pull the plug and have *nothing come in.*

Exactly. Here comes consciousness again. As long as we know what our "rulers" are doing to us and to America, the element of surprise is gone.

Profound and meaningful scientific research will confirm the brain not as the originator of consciousness but as a receiver device. Fig.C1 and Fig.C2 would explain why groups of people can receive identical "hallucinations." Are they hallucinations? Could they be a tear in the fabric of realistic illusions that allows some people to glimpse an equally solid and existing reality that we have been conditioned not to see?

I was amused when I came across an interesting passage in a colleague's book about explaining mind-tricks we can be subjected to. In his book "The Holographic Universe," Best-selling author Michael Talbot deliberated on a hypnotist's performance.

After inducing a trance in a man in the audience, the hypnotist told the subject that he would be utterly invisible to him after being awakened. When the hypnotist awakened the subject, he gazed through his giggling daughter as if she really didn't exist there in person. He also identified the object (a watch) with an inscription wholly hidden from the audience's view, including the subject.

The hypnotized man later acknowledged that his own daughter was *absolutely invisible* to him during the session and that all he saw was the watch in the hypnotist's hand. Talbot explains: "Obviously Tom's perception (hypnotized subject) of the watch was not based on information he was receiving through his five senses." He then elaborated that it might have been a case of mental telepathy. Mental telepathy is but a small subaspect of positronic communication. More to this later.

Furthermore, look into the three-dimensional appearance of people believed to have "died" from the time of Jesus of Nazareth to today. Another example of identical observations would be childhood experiences where (unrelated) children literally "see" identical monsters in their room or closet.

Last, "Near-death-experiences" can be included in this research, although I would instead call them "Near-Life-Experiences." *I will explain later that we are* on the 'dead' side. Just like M. Talbot's example demonstrates, we are tricked by our self-elected magicians into believing realistic illusions into which we can sink our teeth. You can also turn around this sentence and state this: Material physics is the *belief in realistic illusions* as performed by self-styled professional magicians (magicians equals politicians).

With its uncensored knowledge, only universal consciousness deciphers these tricks and erases their solid illusions. In other words, you, too, will discover both the secret of life and the secret that prevented us from seeing it.

Due to the lack of a dualversal model, we are forced to view the above examples as inexplicable, unrelated figments of a "sick" imagination that can be laughed off and ridiculed. Before that, even *educated people of their time laughed off the proposition that Earth might be round or revolve around the sun.*

America has become the testing ground for the widespread manipulation of individual consciousness. If it can be shown that 330 million citizens can be turned into mindless zombies —fearful of blue-clothed authority, electronically conditioned to view an orgy of blood, gore, and killings as mere prime-time entertainment, and illiterate to the point of not understanding election literature, then it would be easy to give the marching orders to impose this mindless system on the rest of the world for a group of elitists diseased by exploding greed.

*Humans without consciousness can be tricked into vot*ing by inculcating a sense of national duty to uphold "moral" values and democracy. What would emerge is a class of professional thieves who shamelessly terrorize the voter in a system of dictatorship that draws its legitimacy from the "mandate of the voters" who helped them gain the majority.

This majority mandate would then justify the creation of laws that would literally allow the extinction of the minority. This is going on right now. Today, we are witnessing the intentional wrecking of our society by Democrats and Republicans alike. Targeted for destruction are children, welfare recipients, low-income workers, teenage mothers, the unemployed, the homeless, the sick, illegal and legal aliens, drug users (but not dealers), gun users (but not dealers), gays, blacks, unbelievers, all people without firm 'moral' character, those who jaywalk, those who spit on the sidewalk, those who cross double yellow, those who eat food or drink water.

Further eliminated will be those who don't steal from the poor, who refuse to snoop on their neighbor, who are not fearful of the wrath of God, all liberals, and those with the audacity to *think*.

Last but not least, all those who refuse their children television time and those who think they can protect owls, trees, or purify drinking water with impunity. Not to forget all criminals who actually breathe and pollute our precious atmosphere with carbon monoxide and carbon dioxide, gases known to contribute to the hothouse effect.

Do you get the message?

It almost sounds comical if it weren't true. Politicians and the media try to convince us that the purge of all undesirables, of all life, except theirs' is the prerequisite for a better society. Isn't it outrageous what we have been conditioned to believe without putting up *any* resistance??

The government forgot one thing on its march toward dictatorship. It can be proven that human consciousness is beyond the reach of prolonged mass manipulation. The farther out in space, the less we can be manipulated.

Whether politicians like it or not, their grip on the human mind quickly ends. Going far out, we could even include the predictions of world-renowned "seers" such as Nostradamus, Edgar Cayce, and others who could not predict events much past 2000. Their foretelling strength was limited to visions of the "electronic" world and could not have (at their position in space and time) included the possibility of positronic awareness with which we will have to deal shortly.

Authority succeeded in blinding our consciousness to the last minute by corrupting all science with royal privileges and money. In return, scientists turned over their research. They provided legitimacy to the fine-tuned machinery of distraction and deception and the presentation of "scientific" out-of-context solutions to non-existing problems.

Take, for instance, the issue of crime and mass hysteria:

In 2023, California spent roughly $20,000 annually per student in the public education system. However, each adult inmate in the prison system costs taxpayers about $132,000. (Locking up a California state prisoner for one year costs nearly twice as much as tuition at the state's top private universities.) Is it wise for the government to reverse these numbers?

The machinery of processing any so-called criminal has become so self-serving, huge, and intimidating to the average citizen. It is comprised of such easy money to steal by revving up emotions that they can't let it go. Childhood is only a tiny passing "phase." Why waste money on it? But uneducated and deceived children without hope of fulfilling the illusion of the "American Dream" will soon rebel.

What is more? The government is channeling alcohol, illegal drugs, and weapons into society to covertly incite young people to violence against each other. Only then can they justify their existence and their mandate to confiscate the collective wealth to maintain "law and order."

Violent and unthinking youth are just right to provide jobs for cops to catch them, judges to condemn them, wardens to lock them up, and probation officers to "supervise" them until they can be seen again, rereleased, and caught again. And so on.

If we complain about the government not providing enough protection from crime, politicians say they need more money to lock up all criminals. This would require higher taxes to pay for more and more police. Up to 500,000 more "troops" behind the thin blue line are in discussion, paid for with all privileges by already starving taxpayers.

Is there much room left to voluntarily pay more taxes when already not enough is left to live a dignified life? Having anticipated this annoying question, the government's top psychologists employ a strategy of emotional blackmail. They pick out randomly, for instance, one case of a child's kidnap and murder, drum it up for weeks in front-page headlines and top television news, and then come back to you with the question: "Isn't it worth a few more dollars to keep our children and the streets safe? How could you be against it?"

Whose heart wouldn't melt when *children* are involved? Certainly not that of the government.

Here, the deception is that a "few more dollars" out of tens of millions of pockets add up to quite some change for the magicians – *but the murderers keep popping up out of nowhere.* Any policy in place is only geared toward catching criminals, including murderers, *after the fact.* Unless the money collected is spent early on *meaningful education to prevent* an otherwise law-abiding citizen from becoming a killer in the first place, it is only being stolen from the public.

In the dirty game of politics and mind control, the government has become so bold that it no longer cares about your opinion. Politicians' master-plan is to get all of your money faster than ever to build as many prisons as they can get away with – for one reason alone: to lock up you and me and anyone else who might object to anything they do.

The California prison population jumped in 40 years (1984/2024) from 23,000 to about 93,500, and 20 more prisons are being planned. In addition to this breathtaking pace, 1,300 to 1,500 bills are signed into law *each year in California alone.* You will soon need a *government permit to rearrange the furniture in your living room* at this speed. Guns, smoking, language, eating, and breathing habits will be regulated to leave no doubt whatsoever *who* is in authority.

Next, constitutionally guaranteed protections will be watered down to complete ineffectiveness. If nothing is done about it, you will soon stand alone, facing a wall of the club-swinging thugs in uniform who can treat you any way they want, including having complete immunity from falsifying evidence against you, drug dealing, rape, torture, and murder. (example: New Orleans and Los Angeles Police departments).

Much of science is on the same side.

The "scientific search" for the physical seat of consciousness is a futile smoke screen. It has the bitter taste of cementing the authoritarian grip on our free will. Only time and further forward motion can liberate us from the delusions of politicians and some scientists who'd like to turn us into zombies in exchange for a Nobel prize.

Creation myths versus creation reality

Since the dawn of mankind, humans have wrestled with the question of their origin. Each society had its own version of the creation myth. Significant is the period in which they were developed: hunter-gatherer societies instinctively felt connected to the natural world.

Bicameral mind societies were more selective in narrowing down the guiding forces to a few crucial spirits and voices they could turn to in time of need. With the onset of human consciousness around 3000 years ago, the human brain could imagine a way to "weed out" all natural views of creation and superimpose on them a single, exclusively responsible deity: God.

Only a conscious mind could ponder the questions of creation, existence, and purpose in a philosophical context. The Greek Philosopher Plato (428 BC–348 BC) is credited with laying the groundwork for political societies.

Essentially, Plato gave all rulers past his time the philosophical blueprint on how to deceive their citizens by establishing the divinity of authority with its 'right' to impose rules for eternal division. "Citizens must be educated to accept the performance of tasks to which they are best suited, but the needed talents are not the individual's choice. Instead, they are determined by the authority." Plato was the first human to discover the deceptive power of language. His thorough explanation of this mechanism gave authorities a powerful tool of oppression. Rulers now knew how to consciously manipulate their less educated citizens by rationalizing the need for the existence of rulers and subordinates.

Having been one of the most advanced thinkers of his time, he also dabbled in creation. He believed the first beings were round entities of such perfection that even God (Zeus) feared them as competitors to his power. Therefore, God (Zeus) cut them into two halves, forcing them to eternally seek the other half in their vain attempt to restore the lost unity.

This ingenious device led the groundwork for the iron grip of Western Religions – Christianity, Judaism, and Islam – on the minds of their followers. Plato's logic figured out that *humans* were the epitome of divine and universal perfection and had the power to challenge the authority of Gods. But he knew that revealing this fact would be disastrous to the upholding of authority that lived on the idea of confiscating the wealth of its productive citizens. So, Plato provided rulers with the techniques to separate the minds of their subordinates from God (cut the perfection of universal beings in half). By explaining the deceptive language, mind distractions, and illusions, chaos and confusion could be instilled in humans to achieve subordination.

This unfortunate technique of public hoaxes is still in place today. It is responsible for the complete division of all people in all societies. The propagation of worth "causes" drives ever more sharpened wedges into the general public, so much so that the ensuing bickering and positioning for winning minor points brings society to demolition.

Today, however, we are witnessing the dawn of a miracle – the switch from human consciousness toward universal consciousness. This inevitable feat is a consequence of universal expansion and brings about an end to artificially created divisions. *Forbidden Cosmology* is a tremendous discovery insofar as it explains all existing elements of creation myths by reducing them to one denominator: *an androgynous, ever-existing universe of the highest intelligence and consciousness that uses the build-dualism to transform itself for the purpose of eternal existence.*

Androgyny is the simultaneous existence of male and female gender characteristics within one body and is evidenced in some plants and animals. Although androgyny is a natural phenomenon, current mind control and demagoguery present it as a "freak" of nature. This removes the possibility of an honest and scientific explanation of the universe. After all, God, or a creator, must be presented as perfect and preferably male in his *dominion* over us.

The "maleness" of God is fiercely defended by religious authorities *and politicians,* for it denotes force and guiding direction – qualities also inherent to secular and religious authorities who want to maintain a system of "wealth extraction."

Discussing creation from a "religious" point of view leads to nowhere. Religion, in general, was a consciously developed product based on Plato's philosophy, which institutionalized widespread deception. The starting point is Plato himself.

Plato's dualism philosophy, in which he maintains God (Zeus) as a separate entity from the first-round beings of God and challenges perfection, underscores the suspicion that he was very close to understanding the dualversal mechanism. What he did was use his genius to obscure it.

It took us over 2500 years to ultimately find the key that unlocks the elements of mythology and replaces it with scientific facts. This key explains the origin of religion, the universe's original state, its expansion, and our function in that system.

Plato's deception was that he bestowed God with the power to mutilate the first round beings of divine perfection (the official introduction of political violence as a tool of ruling). He didn't elaborate on the idea that a God without a perfect counterpart (divine or otherwise) could be bored to death, with nothing noteworthy to do, and thus would not be needed in the first place.

Only the existence of perfectly shaped beings of divine origin next to another divine form made interaction possible. Interaction, however, is not to establish the *authority* of one divine form over another divine form. An honest assessment would include the recognition of an absolute universal equality of divine forces for their mutual and eternal survival. Our continued existence proves this point.

Fig. 3C of this manual presents the scientific version of Plato's idea of dualism, except that it shows who "God" is and why 'he' can survive.

The "perfect round beings of divine origin" *are the cooling spheres* embedded within a superior energy. If the "perfect round beings" were absent in the plasma (plasma as a metaphor for Elementary Antiparticle Energy, Positronum, Originator, or even "God"), it would overheat and self-destruct. By the laws of universal physics, this God could not exist without the mandated counterpart of equal importance: the universal vacuums of cosmic origin.

Only they can act as a host to dangerously high energies through which conversion to electronic matter is possible, much like a womb in biology is the only *natural* way to bring a pregnancy to terms:

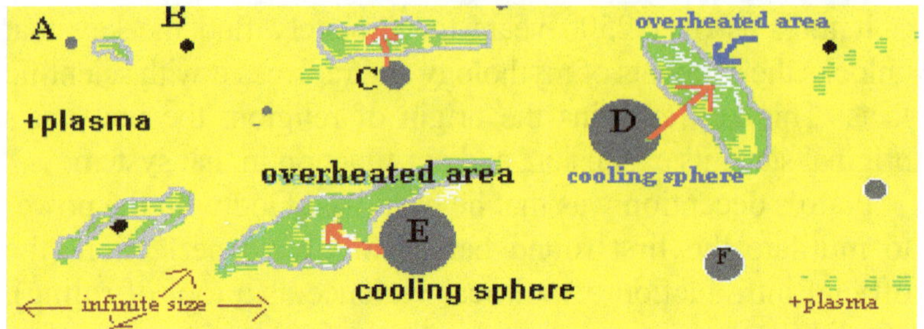

It must be stated that none of these cooling spheres can be "split into halves" *before or during their most functional time.* This is evidenced by the continued existence of our universe, the expanding matter within it, and the verifiable existence of intelligent life, namely us.

This is the ultimate proof of objective cosmic law, don't you agree?

It boils down to this: Humans or other intelligence are not a "freak" accident of nature. We are the direct result of a high-energy conversion. Our consciousness's increasing power enables us to reconnect the brain with the originating energy. This makes us conscious that we are universal beings of the highest intelligence and unlimited power. Bluntly said: we are "I am"! As such, we not only have the power to reject false authority and rigid dogma, we have a universal *obligation.*

Did you ever wonder why today's authorities are implementing the idea of a *"spiritual"* revival of old "values"—the opposite of what we should do? This insanity can only happen because they have conditioned most minds into helplessly accepting dependency and brute force. The mere thought of a universally true independent individual of the highest intelligence shatters their spine.

The most potent tool of oppression is a legislative approach to the laws of physics. By presenting some universal laws of physics and arbitrarily declaring them a result of "biological weakness," as is the case with all gender-relations, they succeeded in declaring them immoral.

Doing so, state and church gained a foothold in regulating biological sexuality to the point of oblivion: 'You can't do it until the relationship is sanctioned, for a fee ("marriage license"). You can't do it with anyone else (lifelong exclusive partner). You can't do it under a certain age (arbitrary legal limit), and you should be ashamed for doing it over a certain age. You can't enjoy it for its plain availability, and you may be "sick" of being proud of it.

Not only do authorities prohibit all genuine and honest expressions of affection, but they also claim to own the power to legislate what humans can do with consenting persons of either gender behind locked doors in the privacy of their homes. At the spur of the moment, they will try to declare all sexuality between persons, including married couples, as illegal acts, which, when enacted, would be the proverbial straw on the camel's back.

All citizens can then be declared hardened criminals with no rights to civil liberties, property, education, or justice.

Why the hardball? Why the upside-down values?

Why is personal affection, true love, and plain, simple nudity made the target of extinction by the Christian Coalition and the so-called "Moral Majority"? Why was it the target of censorship in libraries, magazines, and television? Why aren't drive-by shootings, mayhem, and bloody murders censored?

They instinctively know that love and tenderness lead to discovering the intentionally hidden other half of the universe, which restores our "lost unity." Affection, love, and unhidden honest and universal truth ("nudity") are the supreme qualities of the Dualverse. Or have you ever seen a star in the sky in Gucci garb?

The "Moral Majority" is not the defender of human rights, human dignity, and universal equality. They are screaming because they know they have lost their grip on our consciousness. We are slipping, and thank God, the positronic universe, we are slipping.

We must clarify: personal and universal freedom are not tickets to anarchy where anyone can do anything against everybody. The definition of "anarchy" as a tool of linguistic mind control induces the (wrong) belief that an "absence of government leads to general disorder." The definition *implies* that government is the necessary factor in establishing that order. As we all know, this is not the case. Today's society is beset by chaos, confusion, and anarchy *because of the existence of government.*

If we bring ourselves honestly, we could teach all ages the basic dualversal principle in full consciousness without censorship, obstruction, or profit motif. It would awaken our mental giant to establish

WHAT IF...?

What would happen if we find out that we never die?

What would happen if we were educated with objective intelligence?

What would happen if we realized we don't need space travel to reach other civilizations?

What would happen if we discovered that we don't need doctors or medicine to stay healthy?

... that we don't need schools to get "educated,"

...that we don't need trade schools to "learn" a job,

...that we don't need a brute government to bring order to our lives?

If all of this materialized, our world would have changed from electronic to positronic mode. Human consciousness would be replaced by universal consciousness. We would look back at our current thoughts and be amused at the stubborn primitive model we defended so fiercely to no avail. Honest progress is a universal mandate. If you can't fight it, use it to your advantage.

Can you imagine my surprise when I discovered I had been had? As I developed the manuscript for Forbidden Cosmology, something strange happened: my increasing understanding of the dualversal mechanism opened my consciousness to excitingly unlimited possibilities.

Now I knew why prime-time television peddled UFO mysteries, fake psychics, ghost and angel stories, and near-death experiences like hotcakes. The constant barrage of "mysteries" is intended to keep us confused, amused, and entertained for political and commercial purposes.

No solutions are intended or offered because many of these "mystery" demonstrations are meant to instill false hope in the millions of viewers by believing that the current system is sound and that you could realize the American dream. All you need to do is jump on the bandwagon of glorifying "the Lord," obey authority and send your children to Sunday school.

Intentionally or unintentionally, these rose-colored fairy tales help maintain the division system by suffocating you with many choices. Their range is so broad, covering a seemingly unrelated and wide spectrum, that you are gasping for air in bewilderment: "Please specify your interest of today. We had one UFO sighting over Florida, ten faith healings in Mississippi, two weeping Madonnas in Oregon, two angel appearances in California, one hundred thirty accounts of near-death-experiences, five hundred crystal vibrations in various locations, an albino buffalo born in the Midwest, and we also have on stand-by two hundred professional psychics to help predict your future."

It sounds as if we are still living in medieval times!

. . .

Chapter Thirteen
My Journey to Discovery

If you ever wanted to meet a skeptic of all the above, you have it in me. Still, things happened to me that I can only now understand. Let me explain.

When I found the equation that rules the physics of positronic energy, I didn't understand it immediately. It took me days of head-scratching. It was logical. It made sense. Yet somehow, it couldn't get through to me right away. I felt as if my brain was wrapped tightly in some plastic tarp that kept it disconnected from the rest of my body. I searched my hand-drawn sketch of the dualversal system. I looked and looked.

It hit me like a massive explosion. Wow, I called. Why didn't authorities tell us? I could see the tarp around my brain dissolve and reconnect the brain with the rest of my body. It was an incredible burst in which I could *physically feel the thoughts* rushing forth and back in a never-before-experienced symphony of energy and light. The cloak of electronic mind control and pollution had lifted, and the study of cosmology had led to my liberation.

"Wow," I said again, "that's why they canceled the "SETI" projects. That's why they initially put in orbit an almost blind Hubble telescope. That's why they are raising doubts about and ridicule near-death-experiences. That's why they...oh, gosh!"

At that moment, I became a different person. I could see the artificial and virtual reality imposed on us by education, church, and state.

That night, I watched television, and I couldn't stand it. These newscasters looked straight into your eyes and lied about almost everything. I knew they didn't make the news.

They just read whatever was given to them. I saw the nearly magical circus of politicians gesturing wildly, insulting each other for no reason but to score points for the most stupid act, which we applauded.

We are trained to expect magic tricks and deception and applaud good performance. "Look at this guy. His performance is so convincing—I'll vote for him again!"

That night, I had the best time of my life. I couldn't believe we were still in the Middle Ages, where dancers, elves, and magicians in colorful costumes vied for the deserved applause of spectators and kings alike. I was free to see the performance without the carefully maintained smoke screen of illusion.

The next day, I decided to make copies of my original manuscript. I was elated because I was given the privilege to understand. A deep understanding of the world as it honestly exists, without a screen or deception. It was a *three-dimensional* understanding similar to entering an invisible "bubble" filled with honest and truthful particles that left no room for untruthful ones. Somebody learned about it, and the trip to town was all about that.

When I parked my car in front of the first store on my list, something unexpected happened. Once outside the vehicle, I stepped into ... a different world just like that.

Imagine, I was fifty miles away from home, so it could not have been a case of sleepwalking. It was broad daylight, so it could not have been sleep-paralysis. I was completely sober and not under the influence of any drug, not even Aspirin, so it could not have been a case of hallucination.

Outside my car, time and motion around me slowed to a crawl. Literally, nothing was moving faster than a snail. I was stunned and surprised but not in panic. Entering the store anyway, I witnessed something that gave me goosebumps, making the hair on my arm stand up straight.

It seemed almost sinister how mechanically and automated the few shoppers' movements about the store were. I was vigilant and tried to stay out of their way. The shoppers needed to be conscious of what they were doing or which way to go. With their heads down, they only managed to steal a few suspicious glances here and there—their faces distrustful, old, and drained of energy.

'These are the living dead,' was my thought.

The faster I moved, the slower they became.

When I reached the cashier, I heard the slow and metallic clicking of language, almost as if the recorded sound of a wind chime had been played back at the wrong speed. An automated device thanked me for my purchase. Looking around the store, head high, I couldn't find another human as alive as I was and decided to quickly leave for some badly needed fresh air.

Once outside, I inhaled a never-before-experienced vibrancy from the air around me, became intensely aware of the fragrance of flowers in that air, and watched with amazement a colorful parade of brightly painted cars that crisscrossed before my eyes. These cars were absolutely quiet and ran without any visible exhaust fumes.

I heard some children's playful, happily chirping voices somewhere out of sight. I felt a happiness I had never experienced before. My whole body became a message that expressed only one thought: What an exciting and beautiful world!

After reaching my car, I felt like I was guiding my body from outside. I had the door open, and it showed me how to sit in it and drive away. It was as if I was in that car for the very first time in my life. I had to look at the stick shift to ensure it was the right lever to set it in motion. Another surprise came on the road. The car glided along so perfectly quietly that I thought the pavement was made of cotton candy. Here and there, I saw friendly, smiling people behind the windshields of cars. A silly thought popped into my head. "Isn't this kind of bliss and happiness against the law?" I laughed and started to whistle.

When I reached the next stop, the experience faded out ever so slowly. I shook my head to check if I was dreaming or awake.

The noisy chatter of impatient shoppers reached my ears again, and the fragrance of flowers was replaced by the foul odor of gasoline belching from the exhaust pipes of rusty pickups and rattling sedans.

"Welcome back to the world of the dead," I heard myself saying with disappointment and noticeable distaste.

I had some questions as to the significance of this experience. I wanted an answer and found it in the dual-versal model.

One evening, I looked again at the colored sketch.

"Actually, from the looks of it, I could see that the positronic energy around us could account for what some people claim to be God. It is, after all, extremely potent energy. It exists outside the realm of ordinary matter. The mechanism is of brilliant design, pointing to eternal existence. What is more – we are the converted form of the *same energy*. In this context, we still are "it."

"It" is conscious energy with unlimited intelligence. In our language, this translates as "omniscience." Omniscience is a quality we bestow only on a force we call God. When God equals omniscient positronic energy, and positronic energy is us, then *we are* God, right? Why didn't anyone before me come to the same conclusion?"

Is this perhaps the ultimate scientific explanation of God?

You can imagine that this thought alone would not allow me to fall asleep, period. I turned off the night lamp and stretched out. A couple of minutes later, I heard a familiar voice.

"If anyone could do it I knew it would be you," the female voice said with a tremor of excitement.

Then she stood in my room.

"How did you get in here, Jody?" I asked her matter-of-factly.

"Don't you know where I am?" she asked.

"No," I replied. "Let me in on it, please."

"Don't you know that I 'died' last night?"

"You got to be kidding!" I exclaimed and jolted in a sitting position. "How could you have died when you are talking to me. You are alive! What is going on here, Jody? Is this a joke?"

For some strange reason, I hadn't noticed that the room was strangely illuminated but without causing any shadows. I became aware of this when Jody pointed to my hand-drawn sketch. I could see now, although I remembered I had turned off the night lamp.

"Without your work I wouldn't be here," she said. "I could feel the focus of your thoughts as you concentrated on your sketch. You were the only one I could meet because your mind was free. I still can't believe it myself," she said almost apologetically.

What we talked about that night seemed incredible. I gained knowledge I never knew existed, heard of communication possibilities I could have never imagined on my own, and was assured that intelligence is all around us. The dualversal model was validated from a source that doubled as a first-hand confirmation that we never "die." From this visitor, I heard that universal consciousness just exits a useless or badly damaged biological body without losing its identity, stepping into a reality as accurate as the one we are accustomed to.

I was slightly shaken but wide awake. I had to return to my manuscript. I knew so much, yet I also knew that language was an imperfect and inadequate knowledge translation tool. *How do you explain that* people "die," yet they continue to live? *How do you explain* that I can study hard for months, yet in fifteen minutes of conversation, I can gain more knowledge about everything than learning from all the books in the world? *How do you explain* that we don't need spaceships to reach other electronic intelligence – that all we need is focused thought via positronic energy?

I had to rethink my vision of the world. Did I need another confirmation? Indeed, it came from a very professional and utmost scientific source. A few weeks after my talk with the nightly visitor, I rummaged through some old videotapes I had marked for erasure. I stumbled across a tape about Richard Feynmann, one of the leading theoretical physicists in the United States.

The video showed Feynmann as the lead investigator of the Challenger-shuttle disaster. He was exasperated by his uphill battle with powerful politicians in Congress in his fight to reveal the actual cause of the accident. Congress wanted to hide the truth.

Feynmann insisted on telling it. When the wheels stopped, he threatened to withhold his needed signature on the final report. Feynmann won, and the O-ring story hit the news.

Quickly and mysteriously, the physicist was struck with an absolutely deadly form of a peculiar, fast-growing cancer. His career was cut short before he could succeed in finding the equation for a newly discovered cosmic energy. Daniel Hillis of "Thinking Machines Corporation" worked with him on the project.

Right after Feynmann's death, Hillis found the equation. In the video, he recalls having had a "dream."

Audio transcript, Hillis: "Richard suddenly appeared in my room. I thought I was dreaming. Richard was happy, free, alive, and three-dimensional. I scratched my head in disbelief. We sat down to discuss the project, you know. I told him I had found the equation. Feynmann looked at it and said it was okay. I had known Richard for quite some time, but I wondered if I should tell him the truth, like tell him that he was dead. I knew him well enough as a person who could handle that. I said: "Hey, Richard! It's nice talking to you and stuff. Could I ask you a question? Don't you know that you are, hm, dead?"

"Ah, well," smiled Feynmann, "at least we won't get interrupted this way." (PBS video about the work and life of R. Feynmann, first aired and copyright 1988).

No, this wasn't a psychic show, "Sightings" or "Unsolved Mysteries." It was a scientific documentary about the life and work of one of the greatest physicists in America. The tape slipped through the tight network of censorship that the closed-mind scientific establishment has ingeniously maintained.

<u>Bringing it all together</u>

I now had sufficient evidence to put it together in a larger picture. This has been confirmed: Our perceived reality is an illusion maintained by scientists who are just *believers in illusions presented by professional magicians, e.g., politicians.*

We don't need to go through traumatic stations, such as "Near-Death-Experiences," to see a limited illusion of "heaven, God, or guiding forces." *These things are available right here on Earth! They are around us daily, but we cannot see them under the weight of electronic mind pollution.*

This world works like a parallel reflection of ours. It is fed by the forces of positronic energy and maintained by positron-magnetic field lines. It is divided from us, similar to a vast, angled glass plate whose surface contains an electronically polarized film.

Since we can not open our mind to something we are told does not exist, we can't see this world, just like the hypnotized subject in Michael Talbot's example, explained earlier, *failed to see his own daughter* in front of the hypnotist because he was told that she would be "invisible" to him.

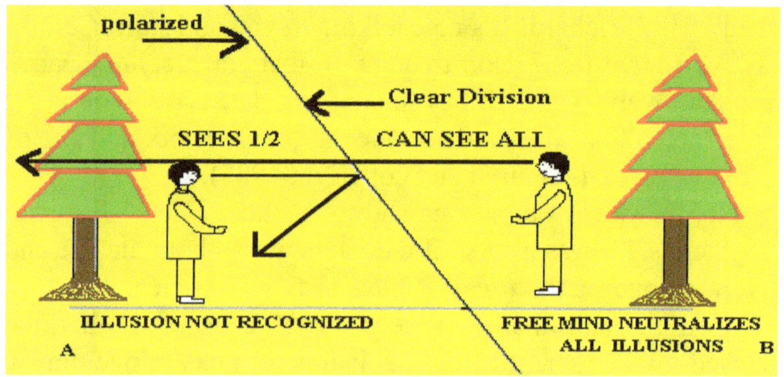

"When you are in the presence of enormous energy, you know it," said author Betty J. Eadie in her bestseller "Embraced By The Light" (C 1992, BJE, Gold Leaf Press, Placerville, CA.). Her experience of heaven is genuine and sincere if one keeps in mind that it is a limited, personal version that anyone else may not repeat. All those millions of readers of her book are trying to solve their own spiritual crisis, and a specific narrow religious version may not satisfy all of these needs.

Forbidden Cosmology explains an objective, much more accommodating universe. It is the original home of all of us—the crystal worshipper, the American Indian, the atheist, the Christian, the agnostic, the Buddhist, the Hindu, the mystic, and the philosopher.

This universe is infinite. This universe is *real.*

There is no need to convince anyone to join a specific group of believers to carve out a larger share. There is no need to worship "a" supreme being when every particle *is "the supreme being."* We will realize that no church is needed to worship this force because this force is you and I, right here on Earth.

We should know and be fully aware that our biological body is nothing less than *the temple of divinity that houses an unspoken secret*: it holds the particle of positronic origin that gives us identity, feelings, and thoughts.

It gives us consciousness, and it acts as a receiver device that wants to reconnect us with "home." Only the magicians of this world, the hypnotists, scientists, and religious dogmatic leaders, are trying to interrupt this connection for greed and power.

The moment we ignore these forces who intentionally blind us with illusions and tricks, our universal consciousness emerges, which immediately reconnects us with positronic energy*:*

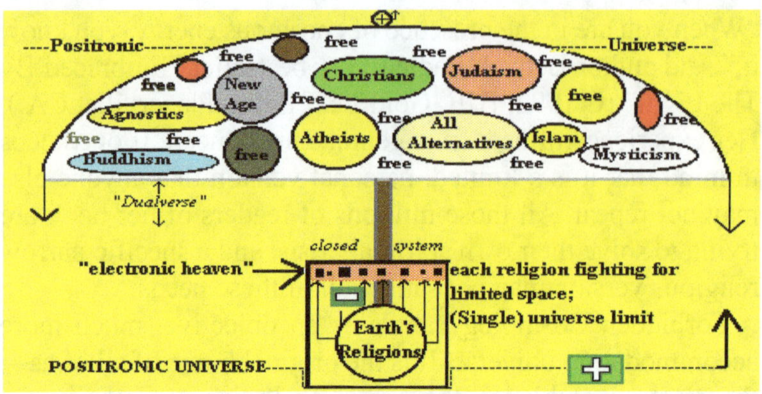

This umbrella above is literally self-explanatory. It shows the limits imposed on us and the solution to overcome them: discard the single universe limit and replace it with *infinite* possibilities.

Just share the joy, the objectivity, the knowledge, and the complete freedom that is available to all. After you experience the connection with your originating force, you know that anything you believe in is just dwarfed *by it. Why would you want to return to the old limits?*

When you ask this question, you have arrived. You may return to this energy at will, without trauma, in total health, from right where you are.

. . .

Positronic Communication

There is no established ritual to communicate with or use in positronic energy. You should have read the complete manuscript, preferably three times, and you should try to understand it. You don't have to concentrate, pray, exercise, beat a drum, sing, light a candle, hold a crystal, or anything else.

You do nothing but look at the dualversal sketch—period! Keep a copy in each of your rooms, one in your locker, and one on your desk at work.

Be aware of each of the three components:

A) our universe at the center,

B) the black vacuum into which we expand and

C) the positronic universe (EE') that surrounds the vacuum.

Touch the positronic side of the sketch and think one thought: "Ah, now I know exactly where it is!" Do this for any length of time– ten seconds, one minute, or ten, and just walk away from it. Do it at least once a day until you become aware that things start to happen that never did happen before. You may walk around "heaven" without immediately recognizing it for a few seconds or minutes. Only on your return to the old world are you reminded of it by the drabness of the colors, for instance.

You may also be visited by someone who had "passed away" but who returns alive and full of energy. When this happens, and you are aware of it, *use it to the maximum*. It will repeat itself.

Again, don't concentrate on it. Just think normally.

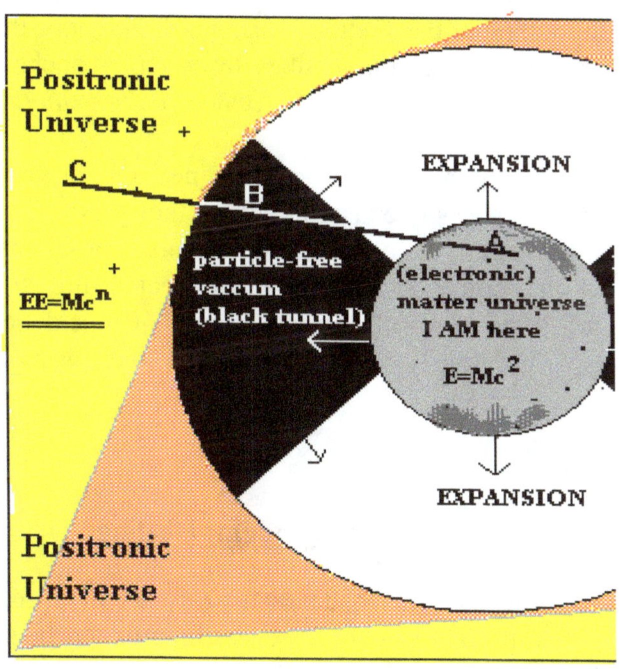

Read the manual several times. With each reading, you will find something new that you overlooked before. Look at the sketch as often as you like. Explain it to your children, or have them explain it to you.

This non-mathematical language of the dualverse is a concept that even a child can understand. Each child *should* understand it. It will be the foundation for all the knowledge to come.

Realize that I did nothing spectacular. I didn't promise heaven to you. *I gave it to you.* It is up to you now to take the key and turn it. Venture into it a little or a lot and be awed anytime by its honesty and objectivity.

If you'd like to solve a problem, it is helpful to *be specific*. In your head, think out loud what you need, and give the positronum a forewarning that you need its knowledge. Say something like: "Please, positronum, I am going to anchor some thoughts in you. Bounce them back to me with new energy and perhaps a solution. I need a new job and more money. Thank you."

Keep looking at the sketch. You can see that your thoughts went through the black tunnel like a laser light. It anchors itself in positronic energy and *remains connected until you find your solution.*

You do this with other thoughts the next day or next week similarly. Soon, a whole laser bundle of thoughts will connect you with objective solutions to personal problems. The answers may come from entirely unexpected directions. Be aware of what you call "coincidences." Coincidences are answers to previously and consciously asked questions that you have forgotten.

. . .

The Power of Thought: Healing

It is incredible how many identical pharmaceutical remedies we have that promise a cure for everything under the sun. Of the 4,500 human diseases, only 20 can be cured.

This means that we are being peddled worthless snake oil in exchange for billions of dollars in prescription and over-the-counter medication. Our mind is purposely conditioned to *do something*, preferably that which works "scientifically." Translation: spend money in exchange for something. When our mind considers it a fair trade, we swallow the pill, *believing* that it will help. In that case, it will help.

But it has been proven time after time that worthless sugar placebos, *when touted as cure-all prescriptions*, worked as well. So, you will pick your nose if you *believe* it will cure your ailment.

Many times, the side effects of several prescription drugs combined are dangerous. Most of these drugs are even poisonous. They truly "cure" tens of thousands of people permanently by completely destroying their immune system and *killing the entire patient*. Side effects from drugs put an estimated 1.3 million people in the hospital every year, and up to 160,000+ of these patients die!

If all artificial pharmaceuticals, prescription or over-the-counter, were pulled off the market, more people would enjoy healthy and long lives. The world's most amazing thing would happen: all those who *thought* they needed the medication to survive would survive *without it and be in much better health.*

Numerous studies have been done on so-called "spontaneous remissions." These are medical miracles in which a verifiably existing and deadly disease disappears entirely and mysteriously.

Is it a mystery?

If we look closer at these cases, we will notice that each patient *did something, and they did it with their minds*. Almost all of them visualized a favorite, cuddly, or otherwise sympathetic tiny monster or soldier cell within their body that had the power to gobble up bad cells. Later, the "bad" cells were removed from the system.

These cures were achieved with a mental focus in an environment of negative (electronic) thoughts, constituting millions of opportunities to interfere. How much more robust, faster, and more numerous would these healings become if all people had conscious knowledge of and, therefore, honest access to positronic energy, in which disease is unknown?

The most notable aspect of two-way communication with this physical energy of utmost purity is that you can push your ailments on it. It may still sound strange to you at this time, but give it a try anyway.

The next time you have a headache or toothache, look at your dual-versal sketch and say, in your thought, clearly and pronounced: "Positronum, please take my (ailment) and neutralize it. I know you can do it. I give you twelve hours to re-energize me and restore my health. I promise to stay connected with you until it disappears. Thank you."

I found this technique very useful. You don't need to repeat or memorize my words as if they were gospel. Just relax and find your own words *while having the model in front of you.*

Our primary goal in life should be, of course, the *prevention* of disease. The here described communication technique works wonders in achieving this. Serious ailments or pain should always be first diagnosed by medical professionals. Afterward, you can put your own mind to it.

Just stay consciously and daily connected with positronic energy. It will work!

The Power of Thought: "Death"

If you are still with me, you already know we can't "die."

We are only hypnotized by earthly magicians with their own agenda. We are the dummies, the zombies, the cannon-fodder who allow them to play a profitable and amusing game. But death is *not a game, especially if no one tells you honestly how it works.*

If you merely look at the medical and biological aspects of dying, we do give up our bodies. In that sense, yes, we have ceased the function of motion and electric (electronic) brain activity. But the positronic function, the universal consciousness of that body, just exits and stands right next to the useless body in the same form and shape it used to be. When this happens, you may stand there awhile and watch the hectic activity around you with amazement. You still possess all of your five senses, plus much more. Are they going to squeeze you back into this body? No, they couldn't. They are much too primitive for that.

When you hear the comment, "He's gone," and they start leaving the room, you wave vehemently and shout: "Hey, wait a minute! I am still here! I am alive and feeling better than ever. My body just wore out. It happens to everybody in this cruel atmosphere and gravity. Look at me *now*!"

But the zombies can neither hear nor see you. They stand before the polarized glass without being aware of the illusion. No electrons, no nothing. Kaput. Finished. Dead.

How wrong they are.

When you find out you can't be seen or heard by them, you may 'loiter' around a little. You notice that nothing whatsoever restraints you.

You are *free*.

At that moment, you are connected to positronic energy.

"Wow!" you say, "what a ride." You see Earth receding, the galaxy receding, and the entire universe. You are home where you came from, and you remember that it was your own choice to go there voluntarily to experience bottled emotions, skin-defined biological restraints, wild gesturing crooks, killing machines, force at all levels, and your will to struggle against it all to make it a lasting experience.

All that was missing down there was *the honest truth*. Everyone seemed to compete for first prize in professional deception. Unselfish honesty was punished by law. What a comical, upside-down world!

Thanks to your newly discovered power, you can now pull the little oddity toward you. You can make it move ever so faster because the closer it comes toward you, the more intelligent you get.

We will only discover that our brain was washed and the mind blinded. Only then will we be determined to upright our world in this *"electronic universe."*

. . .

Chapter Fourteen
Conclusion

I have tried to show that we live in a universe that is only a tiny fraction of the total. I have also demonstrated that attractive forces that shape the universe's large-scale structure must not necessarily result from gravity. I have also shown that our universe had an effective beginning in time and that this beginning was the end of the conversion process from positronic to electronic matter.

The most important aspect of this theorem is the statement that a physical cosmic force exists axiomatically, and a supremely intelligent mechanism is responsible for our existence. This means that the universe is entirely self-contained, but only *electron-ruled matter* has a definite beginning and a predictable end. However, neither its beginning can be viewed as a "birth," nor its end be "death."

It would be more precise to state that the emergence of electron-ruled matter after the Big Conversion explosion is comparable to having gained "independence" from the mother energy. On the other hand, the disappearance of electronic matter is paramount to a "reclamation" process to restore the efficiency of the originating energy. As such, we are always part *of that originating* energy and never die.

This has profound implications for the role of God as we understand it. God can not be the old man with a long white beard who keeps tally of our sins. He also didn't make us in his likeness as our mystic ancestors maintained.

In scientific language, it would be accurate to say that "God," or the positronic energy, made us its *opposite* to ensure its eternal survival. We must now view the laws of physics, as we know them, as something different. In our universe, they are the cosmically mandated set of rules that guarantees that the "energy loan" from the positronic universe will be repaid with interest.

Recreating electronic matter is a mechanism to rejuvenate positronic energy. It also adds to its intelligence.

This physically existing, objective, and intelligent force does not need to answer *our* question of *why* it bothers to exist. It just does.

It couldn't answer our question about its age or whether it has existed for 100 billion or 100 trillion years. Eternal existence means that there is no time frame or reference. Instead, our question *is: What do you know, and how can we get this knowledge?*

Forbidden Cosmology also opens an entirely new chapter of potential benefits. When we can show scientifically that universal consciousness permeates our space and that it can be received by our level of consciousness, we will have found a valid but so far missing link between material and so-called "spiritual" beliefs. This would ultimately unite *all mystic belief systems* and science under one roof.

Whatever a society's or individual's religion or mystic belief is – *it can be projected into the objectivity of the positronic universe.* It will respond in every which way we "pray" to it or approach it, and there would be no change in the results for the traditional faithful, except one: no negativity.

There wouldn't be judgment, condemnation, or imposition, but cosmic justice, universal knowledge, and objectivity. This is not a question of whether we are ready for this. By universal law and using our forward motion in space, positronic energy would force us to open our eyes and minds. This time is about to arrive.

The intuitions and superstitions of our ancestors might come true in the sense that an era of "tribulations" may arrive in which forces of the new universal consciousness will convince dogmatic and superstitious forces to finally upgrade their system to modern time. This process does *not* involve forceful conversion but is achieved by living the example. *Why would you want to stay in the dark if we can show you the benefits of positronic energy,* the benefits of unlimited resources, and abundant knowledge?

This dualism system is so easy to understand and apply that even kindergarteners could grasp its principle in less than 10 minutes. Even *children* would know why we exist because purer energy gives us independence and *needs us* so both of us can survive eternally. We survive in a form called "universal consciousness," which comes from the positronic universe and takes over a biological body to live a short independent life, "a" life span.

When this body is "worn out" by harsh climatic conditions, universal consciousness can exit the body. By returning a lifeless body to the soil from which it was "loaned," a person's universal consciousness detaches from the individual to be in its original state. It becomes absolutely and literally 'unrestricted.' In this form, it can return to the originating energy to refresh itself and eventually return to a *new biological body in which it can experience the sensations of limitations again!*

This mechanism also explains the yet unresolved question of "reincarnation," a belief system that seems to be gaining ever wider popularity.

Universal dualism would also answer the question of whether animals have consciousness. They do. Their body, the 'temple of divinity,' *contains the exact size of a positronic particle* as that which is inside yourself. It has instinct, consciousness, individuality, and knowledge. Animals are only more restricted by the form and shape of their body – chosen to experience its restrictions and limitations on Earth, if only once. Think about this the next time you swat a fly!

The smaller the animals are, the less concerned they are about "death" they are. They know that their bodies (but not their minds) are part of an arranged electronic-biological food chain that enables larger animals to survive longer. Whatever period these animals managed to live naturally, they *instinctively knew* it was the perfect time they could get under these circumstances.

The question of the number of species indicates that biological life is relatively scarce in the electronic universe. Once a place is habitable, trillions of *particles of universal consciousness* will use it to "rent" a biological body of any form to experience its limitations.

The lushness of Earth's nature and its numerous biological varieties underscore that we are one of the few habitable planets in a vast space radius.

Is there a limit to the number and variations of species that Earth could support? No! And no, again. Whatever theories you might have heard of on overpopulation and the threat of starvation are plainly wrong, period. Here is why. At the time of Christ, Earth's population was estimated to be 3 to 5 million. Those guys had a hard time feeding all of their members since their output was commensurate to the productivity of primitive tools. A thousand years later, the world population was estimated to be 50 million, and enough food was produced to feed all of them. Had you asked someone at that time if Earth could ever support 250 million or even one billion people, the answer would have been definite: "Impossible!"

Currently, the world population is a little over eight billion. There is *enough* food available for all of us. The topic of hunger and starvation is not a problem of inadequate food production. Instead, it is a political and profit problem in which rich countries try to maintain their advantage by usurping the entire output of raw and precious materials from poor countries. Unprocessed raw materials are inexpensive, leaving these countries little room for needed investments.

We simply are cowards indulging in total indifference. We view the images of disease and starvation in other parts of the world as part of our daily entertainment. No, I don't mean to be cynical. I am just honest with myself.

This book is a stepping stone to new ways of thinking.

Do something today that will help yourself and others, even if it is just spreading the word about reading Forbidden Cosmology. We need all help making this little blue planet *livable, and we need to do it honestly.*

The world's population could reach 10 or 20 billion. Remember that *each new brain* on this planet is and functions as an *extra receiver* of positronic knowledge. The brain that is not yet born may be the brain that receives *the* most brilliant idea that would allow us to produce food more efficiently and distribute it more objectively. As such, each human life on this planet is divine and *original* and can connect with the *originator* for better and better solutions.

One of these solutions would be rejecting the monetary system, which is responsible for the ills, divisions, and "stratifications" of this world. It was, is, and will remain the false God of this world that distracts us from achieving our universal potential.

The "Golden Calf" stands silent when the music stops.

We will learn a new respect for the originating force, which cannot be corrupted. We have the same divine qualities. If you haven't mastered it, at least *try.*

·   ·   ·

Universal dualism explains many more so-called mysteries straightforwardly and consistently. In this context, we should discuss all possibilities and honestly free our minds from the fascinating times that lie ahead in the new millennium.

"Forbidden Cosmology" is the introductory manifesto of the most basic knowledge *you need today.* Sequels are in the works to further delve into the knowledge I possess in greater detail and in less technical language.

I am sure you will be excited to build on the knowledge

presented here to become the person you indeed are and deserve to be—a universal being of the highest intelligence, positronic origin, and power unmatched in this electronic universe. It is appropriate to celebrate this uniqueness, this love, and this beauty daily. It reminds us of our true *positronic origin and connection with life eternal and profound peace.*

. . .

Glossary

Antiquark: An elementary antimatter particle with an opposite electric charge to that of its matter particle.

Big Bang: The point-the origin of the matter universe out of nothingness, also singularity without symmetry.

Significant Conversion: The process of converting Antimatter to matter.

Big Crunch: The reclamation of all matter.

Biology: Consists of living organisms that repeat the universe's life cycle on a smaller scale but with almost equal power.

Black Hole: The nuclear core remnant of an "attractploded" star, consisting of protons and a layer of electrons. Incoming photons push electrons closer to the nucleus, absorbing their energy. Light "disappears." It is not a source of supergravity.

Chandrasekhar limit: the highest allowed Mass of a cold, stable star above which it must collapse under the force of gravity (1.44 times the size of our sun).

Conservation of energy: The law of physics states that energy or its equivalent in Mass can neither be created nor destroyed.

Electron: Elementary matter particle of negative electric charge orbiting the nucleus of an atom.

Elementary particle: an indivisible particle.

Event: A point in space with an unrepeatable set of coordinates.

Energy, E: Energy can produce motion and heat.

Pure Energy, EE': The capacity to instantly connect in all directions without limit. The universe of the Positron.

Gamma-ray bursts: The penetration of cosmic energy of

higher frequency into the advancing matter universe, deflected by the motion of galaxies and measured as bursts and swirls in all regions of space, including areas without galactic structure.

Gravity is the uncharged consequence of nuclear fusion within a star, which developed due to the electromagnetic inward attraction of charged particles. Gravity disappears when atomic fusion stops, making the star vulnerable to an "attract-plosion."

Intelligence: The power of knowing and understanding.

Infinite Intelligence is the power of instilling purpose in the capacity to know and understand, to which Intelligence must connect or perish by cosmic law.

Light Year: The distance light can travel in one year at a speed of 186,281 miles per second, or 5.89 billion miles annually.

Microwave Background radiation: The light induction of the encompassing cosmic energy of higher frequency on the advancing edge of the matter universe; now so greatly redshifted that it doesn't appear as light, but as microwave radiation.

Near-Death-Experience, "NDE": The rudimentary power of recognition of another, equally important form of cosmic energy with which communication is possible.

Neutron: A proton with a "flipped" electric charge of its internal components (quarks) to avoid repulsion from another proton within nucleon gas clouds in the early universe.

Plasma: the part of the universe that does not consist of void or expanding matter. It is the infinite-sized host of voids and matter with an opposite electric charge to accomplish positional advancement and allow intelligence development.

Photon: A particle of light (massless) moving with the speed of light.

Proton: Positively charged particle. The nucleus of an atom.

Positron: The electron's counterpart, but with a positive electric charge. Elementary antimatter particle.

Pulsar: The nuclear-core remnant of an "attractploded" star, consisting mainly of positively charged protons.

The neutron star spins rapidly, ejecting energy beams when interacting with galactic debris.

Quark: A charged elementary particle. Each of the three quarks is a component of protons and neutrons. It exists also as an elementary antiparticle with an electric charge opposite to matter.

Spectrum: The splitting of an electromagnetic wave into its component frequencies.

Supernova: The explosive "attracting away" of the mantle of a weakening nuclear-fusion star that loses its gravitational strength and can no longer resist the positron-magnetic attraction of the all-encompassing pure energy of the opposite electric charge. It can also happen to a galaxy.

Universe: That part of the cosmos that contains matter and matter particles and expands into a finite-size, gradually diminishing particle-free void.

Universal time: The number of connected coordinates in a continuously accelerating galactic/universal system gives each moving object its time. A certain number of "event-conversions" must occur during consciousness and Intelligence's onset and development.

Void: Cosmic space does not contain matter or matter-energy but contains field forces.

Wavelength: The distance between two adjacent troughs or two adjacent crests in a wave.

Wave/Particle Duality: A quantum mechanical term that says there is no distinction between waves and particles. Both are interchangeable.

. . .

Max Planck's Quantum Principle: Light, or any other wave, can only be emitted or absorbed in discrete quanta in which its energy is proportional to its frequency.

Werner Heisenberg, 1926, Uncertainty Principle: We can never be precisely sure of an elementary particle's position *and* speed: the more accurately we know one, the less accurately we know the other. [To measure a quantum of light, we will need another quantum of light with a different frequency. The higher-energy quantum will disturb the lower-energy quantum to a point where its exact position is uncertain.

Wolfgang Pauli's Exclusion Principle, 1925 (Nobel Prize 1945) states that two similar particles can not exist in the same state; they cannot have the same position and speed within the limits set by the Uncertainty Principle (see above).

All aspects of the "Complete Cosmic Energy Exchange Concept, CCEEC" have been checked against and do not contradict the above-stated principles.

. . .

Bibliography

Ben Turner, (livescience.com) published February 25, 2024: https://www.space.com/ancient-galaxy-upending-cosmology Why atoms are the Universe's greatest miracle,

Eathan Siegel: link: "Do we need a new cosmology?" online article "Earthsky.org/space/a-new-cosmology, 9/5/2023 https://bigthink.com/starts-with-a-bang/atoms-greatest-miracle/ April 12, 2023

Neil deGrasse Tyson, Astrophysics for People in a Hurry, W. W. Norton & Co Inc. NY; 2017

Infinite Possibilities, Mike Dooley, Atria Books, Simon & Schuster, NY; 2009

The Secret, Rhonda Byrne, Atria Books, beyond Words Publishing, 2006

A Brief History of Time, Bantam Books 1988, Stephen W. Hawking

The Holographic Universe, by Michael Talbot, Harper Perennial 1992, Softcover,

The Origin of Consciousness in the Breakdown of the Bicameral Mind, Julian Jaynes, Houghton Mifflin Co, NY, 1976,

The Inflationary Universe, Alan H. Guth, Paul J. Steinhard, Scientific American, Special '91

Particle Accelerators test cosmological theory; David. N. Schramm, Gary Steigman, Scientific American, Special '91

Unified Theories of Elementary Particle Interaction; Steven Weinberg, Scientific American, Spec. 91

Moby Electron; David H. Freedman about Hans Dehmelt, DISCOVER, Magazine of Science, Feb. 91

Wanted: Dark Matter! Marcia Bartusiak, DISCOVER, World of Science, 12/88

The Birth of Everything, J.J. Merritt, Modern Maturity
Magazine, 10/88
What does this...(the atom) have to do with this...(gravity)?
San Francisco Examiner, Spectra Section, D-16, 3/3/91
Sacramento Bee; Various articles by Bee Science writer
Deborah Blum, on mysteries of the universe between 1986
and 1994
On the generalized Theory of Gravitation, A. Einstein
(reprint) Scientific American, Special 1991
 The Discoverers, Daniel J. Boorstein, Random House, 1983
A flaw in the universal mirror; Robert K. Adair, Scientific
American, 2/88
Soul Searching with Francis Crick; by Daniel Voll, OMNI
Magazine, Feb. 94
Quantum Consciousness, D.H. Freedman, Discover Mag.
6/94
Can Science Explain Consciousness? John Horgan, Scientific
American, 7/94

About the Author

G.W. Franke has been a maverick from the start. Born in 1947 in the then "Soviet Occupied Zone" of post-WWII Germany, which 1949 became the German Democratic Republic (DDR), he graduated from High school in 1966. G.W. went on to study languages (Russian/Spanish). He was promptly kicked out of College by the East German Secret Police for the offense of trying to flee the country to West Germany. He worked as a Letterpress/Offset printer to pass some time and attempted to flee the country again. G.W. succeeded in 1972, crossing the Iron Curtain at night, missing a minefield by a mere quarter mile. He studied Law in the BRD (West Germany) for five years and successfully taught himself how to trade commodity futures. G.W. traveled extensively, from Poland to Hungary, Switzerland to the Bahamas, the US to Mexico and Brazil, and more…

A friend called him to the US in 1978, where he supported some businesses while trading commodities. 1980, he got Gold Fever and found his first few ounces in the California Sierra Nevada. In 1983, he became a successful US-registered Commodity Trading Advisor, giving seminars to traders in several US States, from Florida to California.

In 1987, G.W. Franke moved to his current location, where he still owns his property and becomes a landlord. Here, he put the finishing touches to his manuscript 'Forbidden Cosmology,' a revolutionary 'Theory of Everything' that can potentially change our understanding of Cosmology. This groundbreaking theory shifts our Cosmology view from one that started with the inexplicable 'Big Bang' to events before the so-called Big Bang, which led to the BB in an easy-to-understand language. The impact of G.W. Franke's 'Forbidden

Cosmology' theory is profound, offering a new perspective on the universe's origins.

Stars and Cosmology have been on G.W.'s mind since he was 10. At that age, growing up with few resources in a country after WWII, he mused and wondered why the family and country were so poor, lacking goods and resources. Fascinated by the launch of "Sputnik 1" by the Soviet Union on October 4, 1957, this 10-year-old child built the giant flashlight he could get the material for (5 D-cells) and flashed/beamed an SOS ….---… Morse Code Signal into the starry cold nights of winter of 1957. He hoped an alien civilization could notice his signal and come to Earth to help us.

It did just that through an incredible way of new thoughts, presented here.

(The Maverick) G.W. Franke (MENSA Qualified but not a member of Mensa)

Manuscript: "Forbidden Cosmology"

Complete Cosmic Energy Exchange Concept (CCEEC)

Uncovering the Ultimate Connection, August 2024

Other Works by G.W. Franke

Life in "The Positronic Universe."